The Worst Enemy of Science?

Essays in Memory of Paul Feyerabend

EDITED BY
John Preston, Gonzalo Munévar, & David Lamb

New York   Oxford
Oxford University Press
2000

Oxford University Press

Oxford   New York
Athens   Auckland   Bangkok   Bogotá   Buenos Aires   Calcutta
Cape Town   Chennai   Dar es Salaam   Delhi   Florence   Hong Kong   Istanbul
Karachi   Kuala Lumpur   Madrid   Melbourne   Mexico City   Mumbai
Nairobi   Paris   São Paulo   Singapore   Taipei   Tokyo   Toronto   Warsaw

and associated companies in
Berlin   Ibadan

Library of Congress Cataloging-in-Publication Data
The worst enemy of science? : essays in memory of
    Paul Feyerabend / edited by John Preston,
    Gonzalo Munévar, and David Lamb.
        p.   cm.
    Includes bibliographical references and index.
    ISBN 0-19-512874-5
    1. Feyerabend, Paul K., 1924–1994.   I. Feyerabend, Paul K., 1924–1994.
II. Preston, John, 1957– .   III. Munévar, Gonzalo.
IV. Lamb, David, 1942– .
B3240.F484W67   2000
193—dc21   99-13710

100204 8552

T

9 8 7 6 5 4 3 2 1

Printed in the United States of America
on acid-free paper

605884

Preface

I didn't know Paul K. Feyerabend when I came to his Berkeley seminar over twenty-five years ago, about a half a decade before he acquired his extraordinary fame—or notoriety. I had only intended to sit in the seminar, for I had heard that many graduate students were terrified of his critical mind. "What are you going to speak about?" he asked me as soon as he came in and sat down. "I am just sitting in", I answered. "If you want to stay you will have to give a talk," he insisted. "All my ideas are very bizarre," I said. "Par for the course", he said, taking his calendar out. "When will you be discussing them?"

When I did give my talk I experienced in person how disconcerting his criticism could be: it was the sort I would wish on my worst enemy, or on myself if I had really taken seriously the notion that criticism is a main source of progress and improvement. Feyerabend questioned everything, he challenged and sometimes ridiculed even obvious claims. In a conversation with him no idea was taken for granted. That day I tried to dish out as much as I took, though when the session was over I felt so overwhelmed I was sure that in his eyes I must be an obnoxious fool. But he was very friendly, praised my talk, and invited me to lunch at the Golden Bear. It was to be the first of many lunches in which his quick wit would jump from philosophy and science to music, or art, or the theater, and back to philosophy again; the first of many chats in which we would discuss women, or make fun of each other. He was a mesmerizing lecturer and debater. Once engaged by him, it was difficult to notice his crutch and his leg-brace, or to be aware of the constant pain and ill health he had to overcome all through his adult years. He already seemed so much larger than life back then. I remember his animated face, his infectious laughter, and that extraordinary sharpness of mind that delighted me and the many other graduate students and colleagues who became his friends and admirers, several of whom are contributors to this volume, as are others who did not think quite so highly of him.

Paul Feyerabend, who was once described in *Nature* as "the worst enemy of science," was no enemy of science. On the contrary, he showed how complex and exciting science is, and how it may become at once more fruitful and more humane.

Of the many theses that Feyerabend advanced, perhaps the most imporant is that no method, no rule will capture science. Even the most excellent ideas of how science should be done have exceptions. When we examine closely the history of science we see not only that the great scientists violated empirical method, but that they had to violate it—otherwise they would not have achieved the great success for which we now know them.

Moreover, what appears to be the most solid of data may be based on theoretical assumptions, and the difficulties of those assumptions can only be uncovered by looking at the world from a different point of view. Thus even refuted or absurd ideas can prove useful, by providing those different points of view, which, when developed, may bring about revolutions. For example, the idea that the Earth moves had been considered scientifically refuted for almost two thousand years until Copernicus brought it back. To keep science strong, then, we must allow the pursuit of unusual ideas. This means that even non-Western cultures may have important contributions to make to some areas of science. So Feyerabend urged two things: that we do not force science into a philosophical straight jacket (empiricism) and that we do not use the success of science in the West to put down non-Western cultures.

Feyerabend believed that academia had become too conceited and pompous, and so he set out to shock his intellectual audience out of its complacency. At times he called intellectuals "zealots" and "criminals" and much education "brainwashing." And philosophers will tell you that he advocated anarchy in science: "anything goes" (he didn't—he only said that rationalists who insisted on strict adherence to rules would think of his pluralism as anarchy). But behind every outrageous crack there was an intriguing, often a devastating, argument. For example, chapters 3 and 6 of his masterwork, *Against Method*, are among the most brilliant passages ever written in philosphy. Thus to some he was the court jester of philosophy of science, to others one of the most important philosophers of science of this or any other century.

Paul Feyerabend was a man who thought and wrote with the same gusto with which he lived his life. He was born and educated in Vienna, where he studied astronomy and opera before settling for a career in philosophy. It was that career that brought him to Berkeley in 1959 and eventually took him to Zurich. In his final days, in early 1994, with the left side of his body already paralyzed, he penned the last pages of his last book, *Killing Time*, his autobiography. Over the years I had heard about his crazy childhood, the day the Nazis made the Jews sit apart in school, his antics to stay out of combat, the shot that crippled him—the Russian doctors told him that he would never walk again and he daydreamed of rolling his wheelchair down the aisle of an immense library with all the time in the world to read and read. I suspect that *Killing Time* will reveal to a large audience some of the power Paul Feyerabend had to touch the lives of those who knew him. As for us, we will remember him at his best: the light of amusement in his eyes, and the smile always on the verge of mischief.

Gonzalo Munévar

Acknowledgments

The following material has previously been published, and is reprinted here with permission.

"*Sola Experientia*? Feyerabend's Refutation of Classical Empiricism" by Bas van Fraassen, first appeared in *Philosophy of Science*, Supplement to volume 64, no. 4, December 1997.

Part of "Feyerabend Among Popperians, 1948–1978" by John Watkins, first appeared as a section of "Corroboration and the Problem of Content-Comparision" in G. Radnitzky & G. Andersson (eds.), *Progress and Rationality in Science*, (Dordrecht: D. Reidel, 1978).

"Science as Supermarket: 'Post-Modern' Themes in Paul Feyerabend's Later Philosophy of Science" by John Preston, first appeared in *Studies in History and Philosophy of Science*, volume 29, no. 3, 1998.

"Feyerabend, Mill, and Pluralism" by Elisabeth A. Lloyd, first appeared in *Philosophy of Science*, Supplement to volume 64, no. 4, December 1997.

"To Transform the Phenomena: Feyerabend, Proliferation, and Recurrent Neural Networks" by Paul Churchland, first appeared in *Philosophy of Science*, Supplement to volume 64, no. 4, December 1997.

"A *Réhabilitation* of Paul Feyerabend" by Gonzalo Munévar first appeared in his book *Evolution and the Naked Truth* (Aldershot, UK & Brookfield, USA: Ashgate, 1998).

Contents

Contributors

Peter Achinstein is Professor in the Department of Philosophy at Johns Hopkins University, Baltimore, MD. Among his many publications in the philosophy of science are *Concepts of Science* (Baltimore: Johns Hopkins Press, 1968), *Law and Explanation* (Oxford: Clarendon Press, 1971), *The Nature of Explanation* (Oxford: Oxford University Press, 1983) and *Particles and Waves: Historical Essays in the Philosophy of Science* (Oxford: Oxford University Press, 1991). The latter received the Lakatos Award in Philosophy of Science for 1993.

Paul M. Churchland is Professor in the Department of Philosophy at the University of California, San Diego. His publications include *Scientific Realism and the Plasticity of Mind* (Cambridge: Cambridge University Press, 1979), *Matter and Consciousness* (Cambridge, MA: MIT Press, 1984, Revised edition, 1987), *A Neurocomputational Perspective: The Nature of Mind and the Structure of Science* (Cambridge, MA: MIT Press, 1989), and *The Engine of Reason, the Seat of the Soul: A Philosophical Journey into the Brain* (Cambridge, MA: MIT Press, 1995).

J. N. Hattiangadi is Professor of Philosophy and of Natural Science at York University, Toronto, Ontario, Canada. He is the author of *How Is Language Possible?* (La Salle, Illinois: Open Court, 1987), a co-editor of *Aristotle and Contemporary Science, volume 2* (New York: Lang, 1999), and one of the editors of the journal *Philosophy of the Social Sciences*.

Paul Hoyningen-Huene has recently become director of the newly-founded Centre for Philosophy and Ethics of Science at the University of Hanover, Germany. He is the author of *Reconstructing Scientific Revolutions: Thomas S. Kuhn's Philosophy of Science* (Chicago: University of Chicago Press, 1993) and of *Formal Logic: A Philosophical Introduction* (Stuttgart: Reclam, 1998, in German). He has edited and co-edited various books in the area of philosophy of science.

Joachim Jung is the editor of the philosophy journal *Zeitschrift für Philosophie*. He lives in Vienna.

David Lamb is a Reader in Philosophy in the Department of Biomedical Science and Biomedical Ethics, University of Birmingham, England. He is the editor of the Avebury Philosophy of Science series, former editor of *Explorations in Knowledge*, and the author of several books in the philosophy of science.

Elisabeth A. Lloyd is Professor in the Departments of History and Philosophy of Science and of Biology at Indiana University, Bloomington, IN. She is the co-editor (with Evelyn Fox Keller) of *Keywords in Evolutionary Biology* (Cambridge, MA: Harvard University Press, 1992), and author of *The Structure and Confirmation of Evolutionary Theory* (New Jersey: Princeton University Press, 1988, 1994).

Gonzalo Munévar was a student of Paul Feyerabend's during the early 1970s, and is now Visiting Professor in the Department of Philosophy at the University of California at Irvine. His book *Radical Knowledge* was published by Hackett and Avebury Press in 1981. More recently he edited a previous collection of essays on Feyerabend: *Beyond Reason: Essays on the Philosophy of Paul Feyerabend* (Dordrecht: Kluwer, 1991). His latest book is *Evolution and the Naked Truth* (Aldershot: Ashgate, 1998).

John Preston is senior lecturer in the Department of Philosophy at the University of Reading, England. He is the author of *Feyerabend: Philosophy, Science and Society* (Cambridge, UK: Polity Press, 1997), and the editor of Feyerabend's *Philosophical Papers, volume 3: Knowledge, Science and Relativism* (Cambridge & New York: Cambridge University Press, 1999).

Sheldon Reaven studied philosophy of science at Berkeley, where his Ph.D. Advisor was Paul Feyerabend. He now teaches at the State University of New York at Stony Brook, in the Department of Technology and Society, College of Engineering and Applied Sciences, where he directs a graduate program in Environmental and Waste Management, and in the Waste Reduction and Management Institute, Marine Sciences Research Center. He is Executive Editor of *The Journal of Environmental Systems*. He works on conceptual and scientific issues underlying environmental disputes; methodological problems of risk analysis and life-cycle analysis; assessments of new technologies; science and society, and several waste-management, recycling, and pollution-prevention technologies.

Bas van Fraassen is Professor in the Department of Philosophy at Princeton University, New Jersey. As well as co-editing *Current Issues in Quantum Logic* (New York: Plenum Press, 1981), his books include *An Introduction to the Philosophy of Time and Space* (New York: Random House, 1970), *The Scientific Image* (Oxford: Clarendon Press, 1980), *Laws and Symmetry* (Oxford: Clarendon Press, 1989), and *Quantum Mechanics: An Empiricist View* (Oxford: Clarendon Press, 1991).

John Watkins was, until his death, in July 1999 Professor Emeritus in the Department of Philosophy, Logic, and Scientific Method at the London School of Economics and Political Science. Among his many publications are *Hobbes' System of Ideas* (London: Hutchinson, 1965), and *Science and Scepticism* (New Jersey: Princeton University Press, 1984). His last book, *Human Freedom after Darwin*, will be published by Open Court.

Introduction

Paul Feyerabend was not the kind of philosopher whose personality could easily be divorced from his views, either by sympathizers or by opponents. It is therefore no coincidence that several of the essays included in this volume discuss his life and character, as well as his philosophical thought.

Several contributors to this volume were at one time Feyerabend's students or colleagues. Paul Hoyningen-Huene begins his obituary for his friend and former collegue with a capsule biography, detailing both Feyerabend's flamboyant personality, and some of the myths to which it has given rise. He reminds us of Feyerabend's unique position not just as one of the "big four" figures in the philosophy of science during the second half of the twentieth century, but also as an important link between the other three (Karl Popper, Thomas Kuhn, and Imre Lakatos). Feyerabend, despite having a securely scientific background and training, came to take a more hostile stance toward the self-image of modern science than Popper, Kuhn, or Lakatos. He was also tremendously critical of philosophy, especially of twentieth-century philosophy of science. One of his great merits, as both Hoyningen-Huene and John Watkins point out, was the way in which he introduced into philosophical discussion texts from outside academic philosophy. But Hoyningen-Huene is also careful to remind us of some of Feyerabend's own major philosophical contributions: his introduction (alongside Kuhn) of the tremendously influential concept of incommensurability; his arguments for theoretical pluralism; and his critique of a certain kind of self-image that science has acquired, based on the idea that it has a unique grasp on a distinctive *method* of inquiry.

Sheldon Reaven's contribution sets out to sketch the main themes of Feyerabend's autobiography, and of his life. Along the way, Reaven conveys several aspects of Feyerabend's personality that bear on his philosophical stance(s). Problematic relationships pervaded both Feyerabend's personal life and his philosophical orientation during his troubled middle years. During this time, he associated himself with certain philosophical groupings, despite a strong group-aversiveness. But among those who knew him there seems to be some consensus that in his later years he found peace with a new life-partner (his last wife, Grazia), together, perhaps, with a somewhat more stable philosophical position when not associated with any particular philosophical tribe.

Bas van Fraassen's essay is concerned with one of Feyerabend's attempts to refute the 'classical empiricism' of Newton, drawing on a formally parallel argument leveled by Jesuits against the fundamentalist Protestant idea that Holy Scripture can and should be one's *only* guide to faith. Feyerabend's argument purports to show that experience cannot be our only guide to belief, and that tradition must also be taken into account as a separate source of information. Van Fraassen questions whether this response is cogent. He is concerned partly with the scope of the parallel antiempiricist argument that Feyerabend endorses. Does it tell against all empiricist epistemologies, or against all foundationalist ones? Does empiricism still have a place, even after we have renounced foundationalism? Van Fraassen's essay also raises questions such as whether a more Aristotelian empiricism, which conceives experiences as processes undergone rather than as sets of perceptual data (and which Feyerabend seemed to favor in his later work), might escape the argument. And one of the wider issues it broaches is that of Feyerabend's attitude to the enlightenment.

One of the most controversial and, perhaps, most misunderstood ideas developed by Feyerabend is the 'principle of proliferation', which advocates the generation of incommensurable alternatives to current orthodox theories. Feyerabend's appeal to proliferation is closely linked to several notorious slogans, which he frequently endorsed, such as 'anything goes' and his alleged preference for 'epistemological anarchism'. Nevertheless, Feyerabend's thoughts on proliferation are immensely complex and worthy of serious study. This task is carried out in different ways here in the essays by Peter Achinstein, Elisabeth Lloyd, and Paul Churchland.

Peter Achinstein claims that there is something at fault with Feyerabend's appeal to proliferation. The simple invention of a contrary theory without constraints, argues Achinstein, is unlikely to undermine a well-established point of view. Achinstein considers various interpretations of the 'principle of proliferation', draws attention to Newton's 'Rules of Reasoning in Philosophy', and defends Newton's claim that the generation of a hypothesis that is contrary to a universal proposition need not weaken the argument in favor of that universal proposition. Suppose, for example, that it is possible to imagine that there is no such thing as universal gravity, that there is some peculiar grue-type universal force that will be an inverse-square force for another five hundred years, but not thereafter. Such imaginings, says Achinstein, will neither test nor diminish the argument in favor of universal gravity. The mere logical possibility that a conclusion is false is not enough to cast doubt on it if it otherwise has support, he contends.

There is still room for argument over whether Feyerabend's rejections of 'Rationalism' and 'Reason' ("with a capital 'R'", as he would say) are correctly interpreted as rejections of reason *tout court*. John Watkins's essay contrasts with Reaven's in this respect. Reaven's presentation, in his account of Feyerabend's life and work, is compatible with the idea that Feyerabend was constantly trying to widen the domain of reason, thus showing that rationalists such as Popper and his sympathizers had latched on to too narrow a conception of their own ideal. For Watkins, Feyerabend's relationship to Popper's 'critical rationalism' was always problematic, some of his early papers containing misunderstandings of and relatively minor adjustments to Popper's framework, alongside one another. Feyerabend, Watkins notes, found it difficult both to recognize and to acknowledge his debt to Popper. Watkins's account suggests that we should date Feyerabend's disenchantment with critical rationalism to 1967. His subsequent strong reaction against critical rationalism, Watkins believes, went far enough to slip into irrationalism, this being reflected in a decline in Feyerabend's philosophical practice. Watkins's account of both the academic reception for *Against Method*, and of Feyerabend's reaction to it, should be compared with Feyerabend's autobiograpical comments on the same period (*Killing Time*, chapter 12). The core of Watkins's paper, first published during 1978, was written in the immediate aftermath of *Against Method*, a period when Feyerabend's relationships with the critical rationalists were at their most inflamed. Watkins here replies to one of Feyerabend's frequent charges that his book had been seriously misread by one of its reviewers.

Gonzalo Munévar, by contrast, seeks to show that Feyerabend had indeed been misunderstood in certain respects and, in doing so, comprehensively to rehabilitate the reputation of his former teacher. Munévar sees three of Feyerabend's main themes: epistemological anarchism, incommensurability, and relativism, as being firmly grounded in his analysis of the history of science. He shows in some detail how Feyerabend's reading of the case of Galileo challenges both empiricist epistemologies and empiricist methodologies. He also goes to some length to support Feyerabend's claim that arguments from *Against Method* can be successful *reductios* of 'rationalist' positions, which yet do not rebound destructively upon Feyerabend's own stance. But Munévar is also alive to Feyerabend's changes of mind. In particular, he provides one of the first accounts of the way in which, in his last years, Feyerabend disavowed the relativism with which his name is so firmly associated.

John Preston's chapter also deals with certain aspects of Feyerabend's last work, which has as yet been very little discussed in the philosophical literature. Three forms of 'postmodernism' in the philosophy of science are considered: the antireductionism of John Dupré, the Lyotardian rejection of 'metanarratives' and correlative embrace of fragmentation, and the deconstructionist conception of meaning. Preston argues not only that Feyerabend's later work counts as postmodern in all of these senses, but also that his philosophical *practice* is one of the clearest examples of postmodernism we have. However, he suggests that we should endorse only the first of these three forms of postmodernism.

Paul Hoyningen-Huene discusses the relationship between Feyerabend and Thomas Kuhn, with reference to the former's criticism of Kuhn's *The Structure of Scientific Revolutions*. Throughout the '60s numerous commentators saw little difference betwen Kuhn and Feyerabend's respective appraoches to the philosphy of science, choosing to depict them as advocates of irrationalism and 'mob psychology'. In their early works both Kuhn and Feyerabend drew attention to major discrepancies between the actual history of science and the normative accounts that had been developed by philosophers. Independently they both developed nonfoundational philosophies of science, giving expression to the problem of 'incommensurability', which continues to generate controversy. Yet Feyerabend was fiercely critical of Kuhn. Hoyningen-Huene examines Feyerabend's criticism while noting several parallels between the two philosophers, pointing out that in the last decade of his life Feyerabend changed his opinion about his philosophical relationship to Kuhn, and concluding that these influential thinkers were not so dissimilar after all.

The appeal to proliferation is frequently linked to Feyerabend's notorious 'anything goes' slogan, but Elisabeth Lloyd points out that this slogan should not be read as a recommendation for the conduct of scientific research, but rather as a *reductio* against certain forms of rationalism. Lloyd explores the source of this misunderstanding of Feyerabend and suggests that it lies in his defense of the value of a proliferation of views and methods. Now Feyerabend attributed his appeal to proliferation, openness, and tolerance to John Stuart Mill's essay, *On Liberty*. But Lloyd asks whether his appeals to Mill should be read as genuine endorsements of a particular approach or as positions adopted in battles with opponents in order to 'beat them on their own turf', which was one of Feyerabend's favourite strategies. Lloyd supports the first interpretation, maintaining that postfoundational, cultural, and evolutionary pictures of scientific inquiry lend support to the pluralist approach to the pursuit of truth. The problem, however, is just how much tolerance is meaningful? After all, Feyerabend did defend witchcraft, astrology, and several esoteric nonscientific standpoints, and this could be interpreted as a total disregard for empirical evidence. But Lloyd argues that this would be to misunderstand what Feyerabend was actually doing, and even more significantly to misunderstand what he had derived from Mill; namely that blind assumptions regarding the superiority of scientific expertise are not in themselves strong enough to rule out alternative forms of expertise.

Feyerabend's endorsement of relativism for a considerable part of his life and his attempt to locate epistemology in the context of political theory, so argues J. N. Hattiangadi, places him firmly in a tradition that has undermined the intellectual authority of science. Insofar as the liberal democratic tradition has depended upon the external epistemological standards located in scientific inquiry, then relativism with regard to standards of knowledge must pose a threat to the future of liberal democracy. Hattiangadi draws this conclusion from an historical examination of the tension between epistemological relativism and liberal democracy, maintaining that if the latter is possible then there must be standards apart from those communally allowed. But the very existence of those standards is precisely what is rejected by epistemological relativism. So the question posed by Hattiangadi is whether a model of liberal democracy can withstand Feyerabend's version of epistemological relativism. An examination of Sir Karl Popper's analysis of science and his defence of liberal democracy in *The Poverty of Historicism* and *The Open Society and its Enemies* reveals that they are susceptible to Feyerabend's critique.

Hattiangadi concludes with an interpretation of liberal democracy that is compatible with a conception of nonfoundational science.

Paul Churchland's paper also concerns Feyerabend's recommendation of the methodological policy of proliferating competing theories as a means to uncovering new empirical data, and thus as a means to increase the empirical constraints that all theories must confront. Churchland draws attention to our increased understanding of recurrent neural networks to emphasize affinities between Feyerabend's philosophy of science and a 'connectionist' model of cognitive activity. Although he defends Feyerabend's policy as a clear consquence of connectionist models of explanatory understanding and learning, he also criticizes an earlier connectionist 'vindication', and seeks to offer a more realistic and penetrating account, in terms of the computationally plastic cognitive profile displayed by neural networks with a recurrent architecture. According to Churchland, Feyerabend's appeal to proliferation is not to be dismissed as romantic and uncritical pluralism, but is 'a reasoned strategy for enhancing the range of *empirical* criticisms that any theory must face'.

Two weeks before Feyerabend died, Joachim Jung conducted an interview with him at Männedorf Hospital in Switzerland. This last interview, which is included here, covers Feyerabend's thoughts on some of the themes that were central to him: they include freedom and the expression of ideas, ethics, recollections on his perception of the gulf between history and the philosophy of science, and of his differences with Kuhn, Popper, and other philosophers. It is worth noting that in this last interview Feyerabend's passion for argument, revealed throughout his life, was undiminished.

John Preston & David Lamb

The Worst Enemy of Science?

Paul Hoyningen-Huene

(TRANSLATED FROM THE GERMAN BY
ERIC OBERHEIM AND DANIEL SIRTES)

Paul K. Feyerabend

An Obituary

Paul Feyerabend was born on January 13, 1924, in Vienna. He died of an inoperable brain tumor almost a month after his seventieth birthday, on February 11, 1994, in Genolier in the French-speaking part of Switzerland.

Feyerabend was raised in Vienna in less than affluent conditions.[1] He was a very good and inquisitive pupil with many diverse interests. From his early school years, he studied textbooks on university-level mathematics, physics, and astronomy. His lifelong love for music also began in his youth. He played the accordion, took violin lessons, fell in love with opera, sang in a mixed choir, and even took singing lessons in a conservatory. After his high school final exams in the spring of 1942 he was drafted into labor service (Austria was 'unified' with the Third Reich in 1938) and by the end of 1942 he was drafted into the army. After several tours at the front, in January 1945 Feyerabend was hit in the spine by a machine-gun bullet. Thereafter, he was paralyzed from the waist down. After a long time in a wheelchair, he was finally able to walk on crutches. Throughout his life he suffered periodically from intense pain, and he could endure many of his public appearances only under large doses of painkillers, although he never made a big fuss about it. Even in his autobiography, there is almost no mention of the great pains he had to endure.

In 1945 Feyerabend began singing lessons again, at the Weimar Music Academy, and studied theater science. Beginning in 1946, he studied in Vienna: first history and sociology, then physics, mathematics, and astronomy. He also attended courses in other disciplines. In 1948 he visited the Alpbach Forum for the first time. This event takes place annually and includes seminars, lectures, symposia, and cultural events. In the course of his life, Feyerabend visited Alpbach about a dozen times, first as a student, then as a lecturer, and finally three times to head seminars. In 1948 in Alpbach he became acquainted with Karl Popper, with whom at first he became friends, but from whom he turned away in his later years. His interest in the theater also brought him in contact with Bertolt Brecht who offered him a job as a production assistant in Berlin, which Feyerabend declined.

For a long time, Feyerabend said that this was the biggest mistake of his life. Later, because of the kind of people around Brecht, he was no longer so sure. In the late 1940s, Feyerabend founded a philosophical work group together with other students mostly of natural science or engineering. He was perceived as the student speaker. Later the group was named the 'Kraft-Kreis' after its academic leader, Viktor Kraft. Among its invited guest speakers were Elizabeth Anscombe, Georg Henrik von Wright, and Ludwig Wittgenstein. Anscombe gave Feyerabend manuscripts of Wittgenstein's later work. These she discussed with Feyerabend intensively and, according to his own judgment, they had a lasting influence on him. After receiving his diploma in astronomy, he earned his Ph.D. in philosophy in 1951 with a dissertation entitled 'Zur Theorie der Basissätze', which consisted of an elaboration of the discussions of the Kraft-Kreis (this dissertation was never published — a summarized version can be found in the essay 'An Attempt at a Realistic Interpretation of Experience', *Proceedings of the Aristotelian Society* vol. 58, 1958, pp. 143–170). Between 1949 and 1952, Feyerabend traveled to Denmark, Sweden, and Norway where he attended various courses and summer schools. During one of these visits, he became acquainted with Niels Bohr.

In 1952 Feyerabend left Vienna in order to study with Popper in London. He received a scholarship for this from the British Council. Initially, he had intended to study under Wittgenstein in Cambridge, but Wittgenstein died in the meantime. In Popper's lectures and seminars, Feyerabend became familiar with 'falsificationism' and especially with Popper's arguments against the logical positivism of the Vienna Circle. The insight, stressed by Popper (and even earlier by Duhem), that general theories logically contradict the empirical laws approximately derived from them, provided him the following striking argument against the treacherous character of inductivism: how could a theory be inferred inductively from special empirical laws if the theory is in logical contradiction with these laws? Popper's falsificationism, which favored deductive testing, seemed to be the only available alternative to inductivism. Although Feyerabend endorsed Popper's work early on in his career, beginning in the late 1960s he rejected it with vehemence. But his own work concentrated mainly on two other topics, namely, quantum theory and Wittgenstein's philosophy.

In 1953 Feyerabend returned to Vienna after having declined an assistantship with Popper and worked on a set of different projects: he wrote a report (as yet unpublished in its initial form) on the situation of the humanities in Austria after the war, translated Popper's *The Open Society and its Enemies* into English, and wrote several encyclopedia articles. He became an assistant to Arthur Pap for a year. In 1955, Feyerabend was offered a job in Bristol, for which Popper and Erwin Schrödinger wrote him recommendations. Here he began his academic teaching with courses in philosophy of science and the philosophy of quantum mechanics.

In 1958 Feyerabend accepted an invitation to be a guest professor at the University of California in Berkeley. In 1959, he was offered a permanent position there. From Berkeley, which remained his home base until he retired, he had a large impact, especially in Anglo-Saxon countries. He was one of the regular guests of the Minnesota Center for the Philosophy of Science, whose director, Herbert

Feigl, was a friend of Feyerabend's. The center was probably the most important meeting place for philosophers of science at that time. Besides Feigl, Grover Maxwell and Paul Meehl were his most important discussion partners. In Berkeley at the end of the 1950s, he became acquainted with Thomas Kuhn who also had a position there and with whom he soon conversed intensively, and sometimes vehemently. Their main discussion topic was the draft of Kuhn's later famous *The Structure of Scientific Revolutions*, which Kuhn finished in the autumn of 1960. In the 1960s and 1970s, Feyerabend was very famous in the philosophical world. This is apparent from the large amount of (official and unofficial) offers for professorships that he received from Atlanta, Auckland, Berlin, Brighton, Freiburg, Hamburg, Kassel, London, Oxford, and Yale. Feyerabend accepted some of these offers. He taught one or two semesters each in Auckland, Hamburg, Kassel, Brighton, and Yale. For some time, he taught one semester in Berkeley and one in London alternatively, and additionally he commuted from London once a week by airplane to Berlin to teach. In London he became acquainted with Imre Lakatos, who became perhaps the best friend (academically and personally) of his life. (An Italian edition of their correspondence is already published.[2] The English edition is in preparation). Feyerabend turned away from Popper's 'critical rationalism' at the end of the 1960s. Popper became the primary target of biting, sarcastic, and somewhat degrading criticism for the rest of Feyerabend's life, whereas Lakatos tried to connect Kuhn's historical insights with critical rationalism in his methodology of scientific research programs.

Feyerabend experienced the boisterous second half of the 1960s mostly in Berkeley and Berlin. His experiences in Berkeley were crucial for the development of his cultural relativism. As the result of educational politics, more and more members of minorities came to the university. Feyerabend saw the role attached to him as a university teacher as an intellectual imperialist: disregarding the students' backgrounds and providing theses that had no justified claim to generality. Moreover, in Feyerabend's view these theses concerned what a very specific culture, namely that of the white man, perceives as (scientific) rationality: methods that are especially characterized through their abstractness, exemplified through the use of abstract concepts (the critique of abstractness is also the core of *Against Method*, I will return to this later). Accordingly, Feyerabend saw the facilitation of university access for the members of minorities as an act that in no way provided equal opportunities. Moreover, this educational policy secured the predominance of a very specific culture, especially its scientific, technical, political, social, medical, and natural interpretations. A complete test of the presupposed superiority of this culture has never taken place. Feyerabend drew two consequences from this evaluation. On the one hand, he organized his lectures in such a way that the experiences of people from other cultures, and from subcultures of his own culture, could be discussed most authentically. On the other hand, he began to concern himself intensively with the 'rise of rationalism', as he called it, with the creation and dispersion of abstract methods and concepts that began in ancient Greece.

In 1975, the publication of Feyerabend's *Against Method* made him famous far beyond the borders of the philosophy of science. His slogan, 'Anything goes!', became the brand name of his 'sketch of an anarchistic theory of knowledge': the shocking (as he had intended) subtitle of his book. The book was initially to be Feyerabend's part of a book written together with Lakatos, a plan that never came to fruition because of Lakatos's sudden death in 1974. Feyerabend, in his autobiography, called *Against Method* a 'collage' because many different previously written texts, ideas, and arguments were used in it. The main idea of the book was also initiated through a discussion with Carl Friedrich von Weizsäcker in 1965 in Hamburg about the foundations of quantum theory, at which time Feyerabend realized, more than ever, how large was the discrepancy between abstract normative thinking about science (including his own up until then) and the actual, complex, and context-dependent practice of science. This idea remained a central, if not the central, point of his thinking throughout his later life.

From 1980 to 1990 Feyerabend taught the fall term in Berkeley and the summer term at the Eidgenössische Technische Hochschule (ETH) in Zurich. His seminars in Zurich, during which many different topics were quite controversially discussed with guest speakers, drew a large number of participants. After his retirement from both universities, he lead a quite withdrawn life. He worked, on the one hand, on a book entitled *The Conquest of Abundance*, which is concerned with reality and objectivity, but he did not get further than the first few chapters, of which different, often reworked versions have been found. On the other hand, he wrote an autobiography with the title *Killing Time*, which he was able to finish in the hospital while already paralyzed on one whole side of his body, and with death staring him in the face. It was published in 1994 in an Italian translation, and in 1995 in the English original.

If one looks at the course of Feyerabend's life, one is struck by the fact that he was often in places where especially intensive discussions in the philosophy of science occurred at that time. Furthermore, as one of the 'big four' in the philosophy of science in the second half of the twentieth century, he had close relationships with the other three: to Popper in the 1950s, to Kuhn in the early 1960s, and to Lakatos in the 1960s and 1970s. So Feyerabend was informed about the actual developments in the philosophy of science not only on the basis of his readings, but also on the basis of his extraordinarily many personal contacts. Accordingly, many of his works must be understood as reactions to the actual situation in the philosophy of science. Before I turn to acknowledge his work, I would like to comment on Feyerabend's personality. This only concerns a few impressions that are restricted primarily to the years after 1980.

In many ways, Feyerabend was quite unconventional, especially because he did not conform to the established academic customs, sometimes as a conscious provocation. This was because his personal value hierarchy was different in some aspects from the typical or 'average' value hierarchy of academia. In fact, he was always ready to subordinate typical academic values to his own preferences. For example, he revoked his acceptance of a position at the Minnesota Center for the Philosophy of Science, an institution he greatly admired, saying that he could not endure without his singing teacher in San Francisco. He could reject potentially interesting and exciting discussions if a detective mystery that he wanted to see was on television. He refused a meeting with Heidegger that a common acquaintance wanted to arrange just like that. He finished a lecture at the ETH Zurich, where he was presenting himself for the first time, by whirling his scarf like a lasso over his head, then announcing that he had become tired and hungry and so he would go home. At invitations to lectures that he had accepted, the organizers had to be prepared for a cancellation at the last minute (the cancellation might not be because he was not in the mood, but because he was in great pain).

One might think that, as he was moody, undependable, and irresponsible, he was personally an anarchist (as some colleagues put it). Once, Feyerabend justified his last-minute lecture cancellation with the remark that a female critical rationalist argued him into bed — he could not possibly come. His written work was also, with regard to the form of presentation, full of elements that one usually does not find in academic treatises and with which the author signaled that he was a certain distance from the tradition within which he was working. According to one's own viewpoint, those deviations are either amusing or out of place.

Because of his eccentricity, many legends grew up around him. For example, two serious biographical collections claim that Feyerabend had had eleven children[3] — even though he did not have any, and he could not have had any. I do not know the source of this misinformation, but it is conceivable that it was Feyerabend himself, perhaps to show that such things are not especially informative, not to mention relevant. The book *Bluff Your Way in Philosophy* by Jim Hankinson claims that Feyerabend ended his lectures at the London School of Economics by jumping out the window (from the ground floor, luckily), hopping onto a big motorcycle, and taking off with a roar.[4] Not bad for someone whose legs were paralyzed!

From his writing, one can easily get the impression that Feyerabend was arrogant, sometimes even aggressive beyond good taste (the allures of a prima donna). But this impression does not at all match that which one got from meeting him personally. As Feyerabend reports in his autobiography, Carnap also had a bad impression of him from his writing, which then disappeared the first time they met. For many, Feyerabend was an extraordinarily fascinating personality. He could adapt himself excellently to those with whom he conversed, especially when they were his friends. He accomplished this to such a point that from his correspondence one almost gets the impression that there were different people who shared the name 'Feyerabend'. Lakatos is said to have commented on this feature of his friend, "Paul everybody loves you, you have no character". Feyerabend was ironic, full of humor, and always ready to be amiably provocative. Often, he was more interested in the personal circumstances of those he conversed with than in their intellectual achievements. He could be extremely helpful to others, both in institutional and in personal respects, and this was part of the warmth he radiated through his personal magnetism. He tried to be helpful to young people, as well as outsiders, whenever he was asked. His personal independence meant a lot to him. It was one of the roots of his unconventionality. Feyerabend was unreachable by phone because normally he did not answer: only his closest friends knew the ringing code to which he responded. On the other hand, he had a large correspondence; he answered practically every letter, often on handwritten postcards. Letters which he read and answered, he normally threw away, regardless of who had sent them, even if it was a Nobel Prize laureate.

Two things always impressed me a lot about Feyerabend. First of all, he was never conceited. Although he loved both the theater and opera (and as a matter of fact was an actor of sorts himself), he never showed off his enormous reading capacity, his knowledge, his international success, or his intelligence. He seemed to be incorruptible by these things. (Feyerabend often criticized Popper for changing because of his success). For Feyerabend, academic pretensions were loathsome. He dismissed, even rejected, praise of his originality without coquetry. Second, he bore the fate of a wounded war veteran with astonishing calmness, given his paralyzed legs and the terrible pains he endured since the age of twenty-one.

Feyerabend was, and is, one of the most controversial personalities in contemporary philosophy. This is not primarily because of his unconventional personal customs (there are other highly eccentric personalities in contemporary philosophy who are judged a lot less controversial philosophically). Rather, since the mid- or late-1960s, Feyerabend's writings triggered extremely diverse judgments. The reason for this is primarily that they are often highly original, contain sharp arguments, provocative theses, and deep critiques. Moreover, he did this while introducing texts from outside the field, literature that had previously played no role, but which through Feyerabend became fruitful within the philosophy of science. In addition, Feyerabend is, and was, one of the very few philosophers of science who did not bring an unquestioned affirmative relationship to the cognitive achievements of modern science to his work, while at the same time knowing a lot about it. This distance allowed him to see the sciences from different perspectives than the usual ones, and allowed him to say provocative things about them that were not easily swept aside from the start. It seems to me that Feyerabend's strong desire for personal and mental independence, especially from the authority of modern science, is an important key to the crucial aspects of his personality and his work. Even today, his work maintains its potential to excite in several areas of the philosophy of science. On the other hand, in his publications since the mid-1960s, Feyerabend increasingly deviated from the usual (and generally thought of as sensible) academic customs to the extent that his academic reputation suffered. Feyerabend was, of course, completely aware of this: "A recommendation from me could be the kiss of death". This loss of reputation occurred mainly, but not exclusively, with those who did not know him personally: his friends, and those who knew him personally, were prepared to overlook these things as the weaknesses of an extraordinary individual. He was criticized especially for his reaction to the intense critical discussion of his work in the literature, which he perceived as totally inadequate. The general view was that instead of seriously responding to critiques, Feyerabend developed more and more extreme and absurd versions of his views. He exhibited extremely annoyed and sarcastic reactions to criticisms in which, as he believed, he was misunderstood. He sometimes bitched about the authors degradingly and rudely, for example by claiming they were illiterate, or by calling them rodents, or the like.

Feyerabend's reception outside the philosophy of science, which began after the publication of *Against Method* (1975), was also very controversial. There, he established the reputation of 'enfant terrible of the philosophy of science.' In one of the most highly respected scientific journals, *Nature*, Feyerabend was called "the worst enemy of science" (although he was in the company of other alleged science enemies, such as Popper, Lakatos, and Kuhn).[5]

From the very beginning, the great attention that Feyerabend's work received in the philosophy of science was the result of the fact that it contained material that was extraordinarily provocative, independently of his rhetoric. Many of his works attacked presuppositions of the established philosophical tradition that had been more or less taken for granted, and many argued using material from the history of science, or from contemporary science. In any case, one could not sweep aside his works. It was often quite difficult to show that his arguments were wrong, especially at the first attempt, partly because Feyerabend often worked with less well-known material from the history of science or from contemporary science. Feyerabend, in many cases, argued by means of immanent criticism. This means that he took the positions of a certain conversational partner into account, momentarily accepting the presuppositions of that position in order to begin his criticism, without thereby necessarily adopting those presuppositions as his own (although this structure is not always visible, especially in his later works). Feyerabend's affinity for rhetoric, jokes, ironical remarks, insults, and other provocative elements often made the structure of his arguments unclear, sometimes even disguising their substance.

I would like to illustrate the provocative content of Feyerabend's work with one of his most famous essays 'Explanation, Reduction, and Empiricism' from 1962 (many of the same topics were discussed in several other of his essays from the same period). This work became so well-known because Feyerabend there introduced the concept of incommensurability. Incommensurability, in both Kuhn and Feyerabend's versions, continues to play an important role in discussions in the philosophy of science and seemed intimately to have bound together Kuhn and Feyerabend's positions. In this work, Feyerabend attacks both the crucial presuppositions and results that were taken as successes of the accepted tradition in the philosophy of science. His main critical thesis is that Nagel's model of reduction and the Hempel-Oppenheim theory of scientific explanation, which were perceived as shining examples of the fruitfulness of logical analysis in the philosophy of science, fail when applied to "universal theories", by which he meant theories like Aristotelian dynamics, Newtonian mechanics, or quantum mechanics. Feyerabend did not deny that the reduction model and the explanation theory are applicable to empirical generalizations of limited scope, but those, of course, are much less interesting cases. He claimed that the two attempts fail for two reasons. On the one hand, when applied to universal theories, the two models contradict scientific practice. On the other hand, they do not accord with a reasonable empiricism. Thus Feyerabend distinguished between a descriptive and a normative aspect of philosophy of science. Taking these two aspects together shows the sharpness of the critique of the established tradition in the philosophy of science. If claims in the philosophy of science do not match actual scientific practice, this could, in principle, be because of insufficiencies in the practice. However, if the same claims made in the philosophy of science can also be criticized from the perspective of a "reasonable empiricism", that is, judged through a justified normative view, then the critique is devastating.

In the process of criticizing Nagel's model of reduction and the Hempel-Oppenheim scheme of explanation, Feyerabend also attacks other pillars of the established position. Among them is the so-called 'two-language model'. This model assumes that there is a distinction between observation terms and theoretical terms. Accordingly, observation terms acquire meaning independently of theoretical terms. Theoretical terms acquire meaning through observation terms to which they are connected by so-called 'bridge principles'. In the two-language model, meaning flows from observation terms to theoretical terms. Feyerabend attacks this view by claiming that universal theories generate the flow of meaning in the opposite direction. Theories are universal, or all-encompassing conceptualizations of the world that influence the vocabulary that is used in the descriptions of observations, (i.e., observation terms). According to Feyerabend, this aspect of such theories is greatly underestimated in the established positivistic tradition, within which theories are essentially treated as empirical generalizations ("All ravens are black"), that is, just as especially economical summaries of theory-independent facts.

But if one accepts the pervasive character of universal theories, then one must ask oneself how this kind of theory could be empirically tested at all. This gives rise to the suspicion that these theories, by influencing the observation language, exclude the possibility of articulating falsifying observational sentences. There are two usual reactions to this situation: either one denies the empirical character of universal theories, viewing them as valid a priori, or one views them instrumentally, as an aid to the prediction of phenomena, and thus as free from descriptive content. These two reactions are not acceptable to Feyerabend, because they do not fit with his empiricist persuasion according to which theories ought to have empirical content and thus should be empirically testable, and indeed should be subjected to test and abandoned if necessary. Instead of the conventionalist and instrumentalist reactions, Feyerabend suggests that a theory is not, as previously believed, tested by confronting it with empirical data, but that much more serious tests require confronting at least two theories that are incompatible with each other. The weaknesses of a theory often do not appear if the theory is confronted with the facts as seen from its own perspective, but may only appear if facts as seen from the perspective of an alternative theory are allowed. This idea is the core of Feyerabend's view of the necessity of theory proliferation. If it is the case that theories are mainly testable through reciprocal confrontation, then his empiricist persuasion demands that alternative theories should be at one's disposal, thus, his imperative for theory proliferation. Otherwise, there is the threat of dogmatic stagnation. Here, it is of overwhelming importance that the concepts of the competing theories can be mutually exclusive. This exclusion relationship comes about because the application of the concepts of one theory relies on principles that, as seen from the perspective of the other theory, are invalid. This exclusion relationship is called 'incommensurability'. Incommensurability and the demand for theory proliferation were the topics with which Feyerabend had the largest influence in philosophy of science in the 1960s and early 1970s.

Feyerabend connects the special meaning differences that are responsible for the incommensurability of theories with an additional, especially provocative attack on the reigning empiricist tradition. This empiricism demands that the successes of a theory that is succeeded by a theory with a broader scope must be explainable by the competing theory, and that this explanation must fit the Hempel-Oppenheim model. This demand implicitly requires that the concepts of the old theory, as long as they are taken over by the new theory, are taken over unchanged. Thus, the older theory limits the scope of theoretical innovation with respect to which concepts are allowed in the new theory. But according to Feyerabend, the modern empiricist tradition shares exactly this property with "school philosophies" such as Platonism or Cartesianism, and given this similarity, the differences between the school philosophies and the modern empiricist tradition are of subordinate importance. To claim that there is such a similarity is one of the greatest provocations for the later tradition, which began with the pathos of a radical new beginning and a total break with even the remotest rationalistic philosophical position.

The book that made Feyerabend famous far outside the borders of the philosophy of science is *Against Method: Outline of an Anarchistic Theory of Knowledge,* published in 1975. The first English edition was dedicated to Imre Lakatos who was ironically there called a "friend and fellow anarchist". Further reworked editions followed: a revised edition in 1988, and a third edition in 1993. The book was translated into more than a dozen other languages, including several times into Chinese. "Anything Goes" became Feyerabend's trademark, although not at all in the sense that he had intended. *Against Method* seemed to stand for total rulelessness and absolute arbitrariness in science, and thus to exemplify the diversity and equivocation commonly associated with postmodernism. However, this was not at all the idea that the book tried to justify. The target of Feyerabend's attack in *Against Method* was a specific epistemological (self-)understanding of the sciences; one that reduces the special quality of scientific knowledge to the strict application of rules for practicing science. This understanding of science had accompanied modern natural science from the very beginning and, in its essentials, can be traced back to the Greeks of antiquity. Strict rules to achieve a certain target are called "methods". The rules of practicing science are respectively called 'scientific methods', or summarily 'The Scientific Method'. In his book, Feyerabend questioned the existence of such strictly binding scientific methods. Thus the title *Against Method* and its subtitle that contains the concept of anarchism: anarchism as antithesis to the unconditional reign of one or more methods.

The book actually has two parts: a longer theoretical and historical section about the sciences in which the main thesis of the book is justified, and a much shorter section in which the political consequences of the main thesis are drawn. In later books, Feyerabend dedicates himself to the further development of these consequences. The main thesis of *Against Method* claims that science is not an endeavor that is special because of strictly binding methodological instruction, and that it *could not be,* and consequently, *should not be* such an endeavor. This thesis in no way claimed that science is an endeavor in which one can do whatever one pleases, in any way one might happen to feel. Rather, it only claims that it cannot be characterized by absolutely binding rules, like those Descartes specifies in his *Discours de la Méthode.* The existence of methodological instructions in science and also its (limited) success is not denied in any sense. Feyerabend only claimed that such rules in science are not de facto slavishly followed all the time, and that they should not be so followed. There are always situations in which a rule that until now has been fruitful must be broken, if one wants to avoid hindering the progress of science. Soberly formulated, Feyerabend just claims "the limited validity of methodological rules" (the title of an essay that appeared in German in 1972). But how is this relatively moderate view compatible with 'Anything goes'? First of all, one must consider the rhetorical, or more precisely, ironic component of the slogan. 'Anything goes' is an ironic answer to those who insist that there must be absolutely binding rules in the practice of science. Yes, if you insist, says Feyerabend, then I'll give you such a rule, namely, 'Anything goes'! With this, Feyerabend in no way provides incorrect information: indeed, one can discern this as an absolute rule in the practice of science, or in any other practice for that matter, since, being empty, it cannot be broken. The strict validity, independent of the concrete circumstances to which it can be applied, is thus bought at the price of absolute emptiness. Furthermore, when Feyerabend first published the statement 'Anything goes', it came with an ironic footnote about his surprise that people had not noticed that he was joking.[6]

How does Feyerabend justify the limited validity of all methodological prescriptions in the sciences? Rather casually, one finds an abstract justification. The justification is that each methodological rule for increasing knowledge (or for testing or confirming knowledge) is only reasonable relative to certain substantive assumptions about reality and its interaction with the understanding subject. These assumptions are by no means indisputable, but they can change during research, and in fact have changed often enough. Strict adherence to methodological rules thus implies a dogmatization of their underlying substantive assumptions, which of course hinders research and could even, in an extreme case, bring it to stagnation.

Feyerabend puts more weight on the historical justification of his main thesis, especially in the many chapters about Galileo. The idea of the argument is to find, for any suggested methodological rule, an episode in the history of science containing what is generally accepted as an incidence of crucial scientific progress that was only possible by breaking the rule in question. Feyerabend worked through several candidates which seem, prima facie, plausible. To give some examples: that one should not introduce ad hoc hypothesis, that new hypotheses should not be in contradiction with established data or other established theories, that new hypotheses should not have less content than those they replace. He always presents historical examples in which breaking the particular methodological rule in question was essential to the progress of knowledge. From this it follows, as Feyerabend states, following Einstein, that from the perspective of a philosopher of science interested in strict rules, scientists must seem like "unscrupulous opportunists" who, depending on the circumstances, follow or break methodological rules as they please.

How, for Feyerabend, does the main thesis about methodology result in political consequences? Feyerabend thinks that the main cause of the distinguished position of the sciences in an industrial society is the belief in their cognitive superiority in comparison to other forms of knowledge. This belief is based on the idea that science is superior because of its methods, when in fact there are no such methods in the sense of strictly binding rules. Furthermore, the alleged superiority of scientific knowledge over other forms of knowledge has not been thoroughly examined without prejudice in any field. Instead, other forms of knowledge have often been simply swept aside by science. Because of these facts, scientific knowledge has its special social status without justification. It is one form of knowledge among others, which has advantages and disadvantages just like others. Furthermore, it is a lot closer to myth than is commonly assumed. If this is the case, then for a democratic state, the warrant for maintaining a special relationship to science disappears, according to Feyerabend. Just as all religious traditions in a democratic state should have the same rights, all cognitive traditions should receive the same conditions for survival. None of them should be favored over the others by the state. In fact, the special tradition of Western sciences suppresses, without legitimation, alternative traditions. The separation of church and state should be followed by the separation of science and state.

The core of Feyerabend's *Against Method* is a plea against the abstract, especially in the philosophy of science. In Feyerabend's opinion, the abstractions he has in mind do not really lead to universals under which the concrete cases can in fact be subsumed, in a way informative for those cases. Instead, they lead to a misleading watering down and mutilation of the abundance of the concrete. It follows from that that neither normative nor descriptive philosophy of science is possible because they both aim at general norms or descriptions, respectively, of the sciences. In a typically Feyerabendian provocative manner, Feyerabend titled an essay at the Tenth German Congress for Philosophy (1972) 'Philosophy of Science, a Hitherto Unexamined Form of Insanity' because philosophy of science distanced itself from the reality of the sciences, in a cognitively unhealthy way analogous to the loss of reality in some insanities. The program for the political consequences then becomes: "Citizen's initiatives instead of epistemology!" because only a grass roots political process can break the political hegemony of the sciences over other forms of knowledge.

Many of Feyerabend's works after *Against Method* address and develop its main topics. In *Science in a Free Society* (1978), in addition to provocative replies to some critiques, he discusses the political consequences of his anarchistic theory of knowledge. In *Science as Art* (1984), he developed the parallels between the sciences and the arts that are only apparent if one does not conceive of the sciences as methodologically strictly regulated enterprises. In *Farewell to Reason* (1987), Feyerabend continues the development of these main topics. This book's title seemed to confirm the way in which Feyerabend was broadly conceived, that is, as an irrationalist. But the title is mainly a provocation, as Feyerabend in no way intends to wipe out the difference between rationality and irrationality. This can be made visible through the fact that Feyerabend formulates, criticizes, and demands arguments himself, and not only insults, associates, and tells fairytales etc. Furthermore, in certain places Feyerabend explicitly lays claim to reason for himself (and others as well). His point, rather, is to criticize certain theories (namely those portending universality) of reason (and also of morals) and to abandon them. Such portentions, in Feyerabend's views, are tyrannical because they tend to restrict cultural diversity.

In the last years of his life, Feyerabend changed his views about the relativism of cultures, including scientific cultures, and also about incommensurability. The relativism of cultures presupposes that cultures are relatively closed units that have specific procedures and values, and in which another culture should not intervene: every culture has the same worth and has to be respected by other cultures. Incommensurability presupposes or suggests, at least in the eyes of some of its proponents, that the barriers between the different cultures are so great that they are almost totally closed-off from one another. In fact, one sees that cultures often have and have had a vivid exchange in which the most diverse elements of one culture are taken over, more or less unchanged or transformed, from another: there are no insurmountable gaps between cultures in this respect. This suggests the idea that cultures are fundamentally more pliable than is presupposed by both relativism and objectivism (which presupposes the existence of a *single* objective reality): both boil down to cultural essentialism. Directly put, as Feyerabend entitled an essay in 1993, potentially every culture is all cultures. Politically, this has the consequence (among others) that cultures no longer appear as sacrosanct and cannot exclusively be judged from within their own established norms. Rather, their practices can legitimately be evaluated from outside, from a humanitarian perspective.

To conclude, I will mention a further argument that Feyerabend gave for his abandonment of relativism: "If on almost every university toilet door there are relativistic theses, then it's time to distance oneself from relativism".

Paul Feyerabend would probably reprimand me for this obituary because in it, he stands (naturally) at the center of attention in a way which he disliked. Perhaps I would answer that even after Feyerabend's death, I am not ready to follow strictly binding rules about how he should be treated. To this, Feyerabend would probably say, "Yeah, but . . . ," and already we would be into a very stimulating discussion.

Notes

1. I am drawing much of the following biographical information from Feyerabend's autobiography *Killing Time* (Chicago: University of Chicago Press, 1995), and also from *Science in a Free Society* (London: New Left Books, 1978), pp. 107–122. See also John Preston, *Feyerabend: Philosophy, Science and Society* (Cambridge: Polity Press, 1997).

2. M. Motterlini (ed.), I. Lakatos, P.K. Feyerabend, *Sull'Orlo della Scienza: Pro e Contro il Metodo* (Milan: Raffaello Cortina, 1995).

3. *Directory of American Scholars*, 8th ed., *Philosophy, Religion and Law*, Vol. 4, ed. Jacques Cattell Press (New York: Bowker 1982), p. 156; *Thinkers of the Twentieth Century*, 2nd ed., ed. Roland Turner (Chicago: St. James Press), p. 227. I would like to thank Brigitte Uhlemann (University of Konstanz) for her directions to both of these publications.

4. Published by Ravette Books (1985) and further editions, p. 41.

5. T. Theocharis and M. Psimopoulos, 'Where Science has Gone Wrong', *Nature*, vol. 329, 1987, pp. 595–598.

6. 'Against Method: Outline of an Anarchistic Theory of Knowledge', in M. Radner and S. Winokur (eds.), *Analyses of Theories and Methods of Physics and Psychology. Minnesota Studies in the Philosophy of Science*, Vol. 4 (Minneapolis: University of Minnesota Press, 1970), p. 105. I wish to thank Eric Oberheim (University of Hanover) for bringing this footnote to my attention.

2

Sheldon J. Reaven

Time Well Spent

On Paul Feyerabend's Autobiography

Papageno: Wer ich bin? (*Für sich:*) Dumme Frage! (*laut:*) Ein Mensch, wie du. —Wenn ich dich nun fragte, wer du bist? [Papageno: Who am I? (*to himself*) Stupid question! (*loudly*) A man, like you. —How would you like it if I asked you who you are?][1]

One bright and bitter-cold January day in 1996, I stood before the grave of my teacher and friend, Paul K. Feyerabend, in the Southwest Cemetery of Vienna. Partly by chance, partly by design, my wanderings through Vienna led to many of the places that peppered Paul's transfixing stories and embellish his delightful, revealing autobiography, *Killing Time*.[2] In the deep snow by the narrow Feyerabend family plot, I thought of this man's inner odyssey, recounted in his book, from a nonchalant who in this very spot was unmoved at his mother's funeral and who did not bother to attend his father's, to an apostle of embracing tolerance and lovingkindness in his later years.

As each spot in Vienna called up its associations with Paul's stories, I found myself wanting to be able to pick up our conversations where they had been suspended years before. At an enchanting *Zauberflöte* at the Staatsoper, for example, I wished we could schmoose about the Yannis Kokkos sets, designs of such floating, childlike beauty that just seeing them made tears well up. We had sized up many an opera, many a *Magic Flute* in conversations I recall still.

These memories are vivid because Paul Feyerabend was such a mesmerizing, unforgettable presence. So much so that I had always had it in mind to write an article about his personality and his ideas on science for the long-standing *Reader's Digest* feature 'My Most Unforgettable Character'. Paul whooped merrily the one time I mentioned this scheme to him. With the same impishness with which he put his astrology chart on the author's credentials flap of the dust jacket of his *Against Method*, Feyerabend savored the very idea of publishing such a story in the *Reader's Digest*: not in some scholarly dustbin, but a journal—no, a mere *magazine*—disdained as the height of lowbrow in intellectual circles.

There is a more telling reason why *Reader's Digest* would have been a fitting forum, beyond the chance it offered to tweak academic vainglory. Feyerabend

16

liked the *Reader's Digest* because it was not specialist turf. Here indeed was one of the very few general-interest publications left. Each issue had a little of this, a little of that—history and biography; science, technology, and their impacts; art and politics; practical household tips and dramatic stories; jokes. This animated mix was just Feyerabend's cup of tea and is exactly what one finds in *Killing Time*. Reading the book is like spending an evening with the author himself, cruising through many subjects—philosophy, women, science, jokes, theater, ancient history, women, music . . . Feyerabend also liked the *Reader's Digest* because it was not heavy-handed: he often seized life in full measure, yet advocated and displayed a light touch in one's doings. As his autobiography makes clear, this buoyant stance was in some respects a mask, a courageous show.

Feyerabend barely managed to complete *Killing Time*. He died seconds after hearing his wife Grazia whisper the happy news that its publication was at hand. One now has to imagine Feyerabend's ebullient voice reading his words aloud. In this dramaturgical sense, *Killing Time* is for me something like a condensed staging of his life. During a busy lifetime he did a lot, read a lot, saw a lot, thought a lot, and as usual, he has lots of interesting things to say about what he did, read, saw, and thought. So we learn something about where he got his ideas about science and philosophy and life, how they developed, and what he came to think about it all in his last years. And we learn a great deal about his family; his experiences of the Nazi era; his physical ailments; his dream life; his pinball-machine career; his friends and lovers and wives; his favorite singers, actors, conductors, and authors; and his most treasured individual opera and play performances. Parts of *Killing Time* read like a Baedeker of intellectual and cultural gossip (which Paul enjoyed thoroughly): many juicy details are provided and many famous names are dropped.

Underneath these particulars, as I see it, run the real main themes of *Killing Time*. These themes are *courage, human credulousness, living* con brio, and *loving-kindness*. Countercurrents to these themes also are recorded in the book, sometimes with confessional frankness. Feyerabend's Viennese warmth, charm, and sparkle were not the whole story, anymore than is the case with Vienna's own history. We're talking about a human being here, one who had to face tremendous odds at that, and not a saint. Because of its mix of human virtues and failings, in *Killing Time* Feyerabend has left us an intriguing mystery (Feyerabend loved them), as well as a compelling personal and intellectual tale.

Courage

> Papageno: Bleib zurück, sag ich, und traue mir nicht; denn ich habe Riesenkraft. (*Für sich*). Wenn er sich nicht bald von mir schrecken lässt, so lauf ich davon.
> [Papageno: Stay back, I say, and trust me not, for I have a giant's strength. (*To himself*). If he doesn't go away scared soon, I'll have to run for it.]

Feyerabend showed great physical and intellectual courage throughout his adult life: the former because he had to—in combat, in battling painful infirmity, in facing cancer—the latter because he chose to, because it was in his nature so to choose. His formative years did little to prepare him for these ordeals. We read of Feyerabend's family origins and of his upbringing in rather modest and often unruly and confining circumstances. The unruliness seemed unremarkable at the time, a given among the urban lower middle class, and the restrictiveness suited the already inveterate reader. He reviews with bemusement his high-school teachers and curriculum, especially his early studies in physics, astronomy, philosophy, and singing.

Killing Time portrays the social atmosphere at the time of the Anschluss, when Paul was fourteen. His father was a Nazi party member. The teenage Feyerabend at times was enamored of the Hitler Youth, and thumbed his nose at them at other times. In both cases his motives were aesthetic (e.g., cool uniforms) and not ideological. He recalls that Hitler's speeches were not the full-tilt hysterics excerpted in documentaries, but were just as dramatic for their contrasting soft voices, carefully and deliberately building up to the hysterics, which were anything but a loss of control. Feyerabend saw how Jewish professionals and civil servants were fired and schoolchildren were ostracized, and how all Jews were forced to wear yellow stars before they began to 'disappear'. He seems to have neither endorsed nor been taken aback by these measures. At one point, Feyerabend toyed with Rosenberg's racial and cultural nationalisms; two years later, they left him cold.

> As for myself—I certainly cannot undo my wavering and unconcern during the Nazi period. Nor do I think that I can be held responsible for my behavior. Responsibility assumes that we know the alternatives, that we know how to choose from among them, and that we use this knowledge to push them aside through cowardice, opportunism, or ideological favor. But I can report what I thought and did, what I think about these thoughts and actions today, and why I changed. (p. 175)

Each reader must judge for himself what standards of *moral* courage and insight are to be demanded of an impressionable youth who turned twenty-one in 1945. *Killing Time* contains several interesting discussions about evil, compassion, and anti-Semitism, and how in the postwar decades Feyerabend ever more viscerally felt the monstrousness of intolerance. During the war, though, Feyerabend was stationed in France, Austria, Yugoslavia, Russia, and Poland. He briefly toyed with the idea of joining the SS, but ended up in the regular army, where he became an officer because the schooling itself meant time away from the front. Told in 1943 that his mentally long-suffering mother had just committed suicide, Feyerabend recalls, "I felt absolutely nothing"; at the funeral, "people commented on how cold and unmoved I looked throughout the ceremony". It took Feyerabend many years to overcome this self-absorbed, cold-fish attitude.

He was decorated for bravery on the Russian front, as a combat engineer lieutenant effectively in charge of several thousand soldiers. Like many veterans, he is dismissive of his behavior in the fog of war: "During battle I often forgot to take cover. It was not out of bravery—I am a great coward and easily frightened—but out of excitement . . . it was like a theater and I acted accordingly". He saw untold suffering and witnessed two atrocities (minor ones in the context of the Second World War) that made him shiver only in later years. Many war memories are vague or simply blank. Today this might be attributed to post-traumatic stress syndrome.

Feyerabend was twenty or twenty-one when the bullets came that crippled him for life. From then on he often was beset with crushing pain. I will never forget how Paul would wince from and be drained by this cross, how he was driven to a succession of herbalists, acupuncturists, and massagers in his search for relief. Equally trying was the mental anguish of his impotence, a further result of the bullets. *Killing Time* recounts the adaptations and stratagems Feyerabend made use of in his relations with women. He makes light of this, but the inner devastation of such a blow to this handsome charmer surely lies behind much of the roving, rejecting, fleeing patterns in his relations with women when he was single, and with his first three wives.

Restored to Vienna after the war, Feyerabend expected to concentrate on physics, mathematics, astronomy, acting, and singing; he had a beautiful, ringing "world voice". *Killing Time* inventories the lecture subjects and personalities of his teachers, often pointing out how a memorable art history class, for instance, bore fruit in some Feyerabend idea or project thirty years later. This postwar period was for Feyerabend a happenstance of grabbed and missed opportunities. For example, he turned down a chance to work with Brecht. For years, he saw this as his life's biggest mistake, but came to realize it was a blessing in disguise to have avoided Brecht's personal obnoxiousness and the stifling orthodoxies of his camp followers.

Feyerabend became active in the Alpbach summer conferences, in which stew one could find Meitner, Kreisky, Koestler, Krenek, Dirac, Kmentt, von Hayek, Bertalanffy, Hollitscher—and the comedian Helmut Qualtinger, whose hilarious routines happily are still a mainstay on Air Austria flights—and Karl Popper. These close encounters are engagingly encapsulated in *Killing Time*. Feyerabend's big breaks came from Alpbach and, in the form of Anscombe and Wittgenstein, through the Kraft Circle (Feyerabend was its student chairman). These led to postdoctoral study in England with Popper, the rise and fall of which association is laid out in the book. Feyerabend, back in Vienna, assayed several academic and nonacademic prospects. He "loused up" his professional operatic opportunity. This was gallingly disappointing to this superb singer. In something of an afterthought, he applied for his first professorial job, at the University of Bristol. So Feyerabend's release into the warrens of academic philosophy came about circuitously, almost inadvertently.

Throughout this postwar period, Feyerabend's reputation grew as a bold, outspoken *enfant terrible*, a hovering gadfly, at Alpbach, in the Kraft Circle, and in several other intellectual and artistic venues in Vienna and England. A spectacularly apt characterization of this side of Paul's personality was made by my father in the 1970s, who wanted to know who my dissertation advisor was, this guy Feyerabend I kept talking about. After my long-winded exposition, my father humorously sized up Feyerabend as a "Moishe kaperach"—that is, a "contrary Moses":

whatever you say, he'll likely argue the opposite. This streak was partly fun, partly Socratic instruction.

I believe that the intellectual courage it took to challenge many celebrated ideas so directly seemed to Feyerabend as nothing by comparison to the courage he had had to summon in order to persevere in war and infirmity. What could the academic world could do to him—turn him down for a job, deny him tenure, call him names? *Killing Time* tracks this devil-may-care (on the surface) restlessness through Feyerabend's sometimes simultaneous positions at Bristol, Berkeley, New Haven, Kassel, Berlin, Auckland, Brighton, and Zurich, driven by desires to escape boredom or stuffy colleagues, seek peace and quiet in some periods or cultural excitement in others, and get comfy pay for light teaching loads.

Feyerabend's personal and intellectual experiences in these cities, and his artistic and romantic adventures while visiting London, New York, and other cultural entrepôts, make up perhaps half of *Killing Time*. The cast of the innumerable vignettes includes most well-known philosophers of science, dozens of famous singers and actors, many prominent scientists, and celebrities in other lines of work.

Paul Feyerabend admired courage and boldness in all their forms, and whether in philosophical, musical, or literary dress. He especially respected well-informed opponents (not lazy "illiterates") who had the courage of their convictions, who advanced them with enthusiasm and passion. Nothing would be more distasteful to him than today's manifestations of a hip, smug, knowing distancing from, and cynicism toward, great traditions in scientific thought. Now it seems that many have taken up the politically correct gospel, often laying it at Feyerabend's feet, that scientific theories are deconstructable, arbitrary, relative (more on this below), and that those in the know are too cool, too smart, too cynical to embrace them or take them seriously. Maybe (say) Galileo had some rhetorical tricks up his sleeve, maybe his opponents had much stronger arguments than often supposed. But for anyone to dismiss Galileo—a towering scientist, consummate writer, and heroic soul to Feyerabend—on those Feyerabendian accounts would incense him. You see, Paul was not a cynic at all (let alone a relativist concerning truth).

Feyerabend's courage did not mean an absence of fear. Far from it. A massage therapist in London "seemed to notice the vast amounts of fear I carried around with me". A 1967 notebook entry reads, "So one day passes after another and it is not clear why one should live". Feyerabend comments, "Sentiments such as this have been faithful companions in my adventures".

Courage was marshaled a final time as Feyerabend faced brain cancer and prepared to die. His meditations (I use this word because Paul so treasured Marcus Aurelius) on compassion, friendship, and above all love, on having to leave his great love Grazia, are deeply moving.

Human Credulousness

> Papageno: Hm! hm! hm! hm!hm!hm! hm! hm! hm!hm!hm! hm! hm!
> hm!hm!hm! hm!hm!hm! hm!

Killing Time chronicles in bits and pieces the development of Feyerabend's thinking on the nature of science and its place in society. Here is a page or two on the
origins of an idea, there a protestation of criticisms once or now in currency, here
a reaction to this or that subsequent development, there again a discussion of
what, in hindsight, should be reformulated. For many readers, these nuggets will
come as embroidery, since almost all of them have been presented at some length
in Feyerabend's other books; this one is an autobiographical sketch, not a disquisi-
tion. Yet the intellectual panorama in *Killing Time* does suggest a metaphorical
appreciation of some Feyerabend leitmotifs.

I think that what he saw in his theoretical and historical studies and noticed
in himself and others was that *everyone is credulous, people will believe almost
anything*. The term *credulousness* is meant to mark a neutral place lying between
human fallibility, on the one hand, with its connotation that while it's always
possible that this or that belief is erroneous, we can fairly handily extricate ourselves from any single such predicament, and human gullibility, on the other,
with its pejorative connotation of being suckered, its suggestion that the one deploying the term is exempt from it. People can be induced to believe in many
ways: wishful thinking, intimidation, self-aggrandizement, emotion, overcaution
and by what they see, by the limited instruments of reason, by not being aware of
alternative views.

The learned are no exception in this regard. One of Feyerabend's great contributions was to show that scientists, for example, are more swayable *by reason*, or
at any rate what people call reason, than is good for science.

Imagine someone reading Plato for the first time, coming away convinced by
the indeed elegant arguments, the brilliant examples, the replies to objections.
The same person, upon later reading Aristotle, will wonder how he ever could
have fallen for Plato—in light of Aristotle's elegant arguments, etc. The person
under Plato's thrall does not even see what a fundamentally different point of view
might be like. If one views Kant as the next fully blown alternative to both Plato
and Aristotle, one can see that it can take a long time to bring new views to life.

Feyerabend's study of celebrated episodes of theory change in science found
broadly similar forces of boldness and tenacity at work. Arguments and experiments that seemed decisive were not; there were jumps in argumentation; objections were swept under rugs; many phenomena and experiments could be described, interpreted, and explained in entirely new ways. None of this was bad; on
the contrary, all of this made up the engine of progress. Physical science, arguably
the most powerful edifice of reason (using the term broadly), was a much more
ramshackle affair than it seemed.

Because of the intrinsic difficulty of the problems science poses, spotting these
holes in reasoning, and creating new theoretical frameworks, was, and is, extremely hard for even the best minds. No wonder that scientists were too easily
swayed, or that sometimes they refused to be swayed, out of habit, or laziness, or
inability to see or entertain the significance of objections, or by seemingly powerful counterarguments, or by imaginative interpretations of phenomena.

Many observations of this sort led Feyerabend to a more general appreciation
of reason as a double-edged scalpel that both uncovers and conceals, a tool of

discovery and deceit. If reason showed that it cannot always be trusted, what then? This is not the place to assess the many paradoxes, real or supposed, in this view of reason — for example, does it mean that "anything goes" epistemologically? Is it an intelligible view to begin with? How can it be criticized, since objections and contradictions are themselves instruments of reason? This territory is much traveled elsewhere. Feyerabend devoted a lot of energy to maintaining that his views on reason and incommensurability were not self-refuting, *and* that such a demonstration was not needed in the first place, *and* that all three terms were themselves murky—which made them conceptually rich, fruitful. Feyerabend refers to "reason—whatever that is".

He says in *Killing Time* that he never denigrated reason, and my own view is that it is quite, well, reasonable to read much of Feyerabend's thinking on this topic as derived from the view that even the best scientific reasoning is never, or rarely *airtight*, that in fact it is often *full of holes*, but that it nevertheless is one of humanity's towering achievements, a monument of, by, and for "reason". Suffice it to say that Feyerabend saw reason as a tool in the way Churchill saw democracy as a form of government: the worst one, except for all the others. Its judicious use is an art, not a cudgel or a recipe.

Feyerabend was well aware how easily he, too, could be persuaded by what seemed to be pretty good arguments. In some respects he found this a problem, in other ways a gift. Several places in *Killing Time* describe his awareness of how his views would seesaw, pulled one way then another by brilliant . . . reasoning. As he puts it, "Moreover, I could be easily convinced of the merits of almost any view". Lacking an ironcast view, he often floundered about. Nevertheless, he considered himself fortunate to lack such a view. He also said many times that he would switch sides if one side became too dogmatic, gave in too easily, grew unduly powerful politically. That is partly why Feyerabend argued for the separation of state and science, and why he thought Galileo deserved defense even if Galileo's arguments (reasoning?) were not what they were cracked up to be.

Besides, Feyerabend never claimed that his own views were immune to his own views. He almost seemed to dare others to come up with solid, well-informed, *well-reasoned* (even though there is no recipe or test for this trait) counterarguments, and even an entirely new supplanting picture of the nature of science. Feyerabend always was the first to acknowledge that his views on scientific 'swayability', on the handwaving and telling brilliance found simultaneously in the best edifices of reason, applied equally to *everything he himself was saying or writing*. This is why the cunning of reason always was a Cheshire cat affair for Feyerabend. Again, much has been written about whether or how he "solved" this question.

To *combat* credulousness, strenuous effort is required. One tool is *having many theories* competing; Feyerabend here was inspired by Mill. A second tool is *hard work* before opening one's mouth. Although *Killing Time* often portrays its author as a professional lazybones, Feyerabend often was a workaholic and thoroughly mastered his subject material and its historical antecedents. He always reserved his most withering criticism for unprepared 'illiterates' who have not done their research. As said earlier, Feyerabend appreciated having worthy opponents. The third Feyerabend weapon against credulousness is a skeptical attitude toward authority. Without these protections, intellectuals can fall victim to all sorts of fads, such as deconstructionism—"whatever that is". This brings to mind Orwell's pronouncement: "You have to be an intellectual to believe such nonsense. No ordinary man could be such a fool".

Paul Feyerabend loved to gore the oxen of academic martinets. He was from the start the bête noire of the Beckmessers, Dr. Fossils, and Major General Stanleys of learned circles. I am sure he would place today's politically correct wardens in the aforementioned company. Take this riposte, for instance:

> The early critics [of *Against Method*] were rationalists and science freaks. Times have changed, and so have the standards of political correctness, but chauvinism, illiteracy, and intolerance are still with us. An example will show what I mean. "Any reader of Feyerabend," writes Hilary Rose, "must see that his philosophical prescription of 'anything goes' is profoundly linked to his lewd sexist conception of a new theory as a charming courtesan whose sole purpose is his delectation . . . It goes without saying that the 'you' who 'can do anything you like' is profoundly gendered. No one could for a moment consider that women were being invited to do anything we like". Well, I am accustomed to weird remarks, but this one certainly takes the cake. "Is the woman nuts?" I exclaimed when reading the passage. "How on earth did she get the idea that AM is for males only, and even 'profoundly' so?". (pp. 148–149)

Another disparagement shows how he was appalled by the politically correct segments of the 'Feyerabend industry':

> There are many ways of thinking and living. A pluralism of this kind was once called irrational and was expelled from decent society. In the meantime it has become the fashion. This vogue did not make pluralism better or more humane; it made it trivial and, in the hands of its more learned defenders, scholastic. . . . People, intellectuals especially, seem unable to be content with a little more freedom, a little more happiness, a little more light. Perceiving a small advantage, they seize it, circumscribe it, nail it down, and in this way prepare a New Age of ignorance, darkness, and slavery. (p. 164)

The most troubling aspect of Paul's legacy is that he is regularly accused or credited, depending on who is talking, of having provided the ideological heavy armamentarium for 'antiscienceism,' deconstructionism, other modernisms, and every 'anything goes' theory (UFOs, etc). This charge, or commendation, is, I think, unwarranted. Feyerabend's displeasure with those who take a dismissive, cynical attitude toward the great traditions in scientific thought has already been mentioned. More importantly, nothing pervaded his writings and his personal life more than his fight against *intolerance*. He feared and despised it in its ideological, political, and intellectual embodiments. His target always was the "-ists"—Marxists, history revisionists, Popperists. If there are today Feyerabendists, he would be the first to ridicule and disown them.

Living Con Brio

> Papageno: Der Vogelfänger bin ich ja, stets lustig heisa hopsasa!
> [Papageno: The birdcatcher, that's me—always merry, hi-de-ho!]

Paul loved to live. Many periods of his life were a flurry of dinners, plays, operas, romantic evenings, wrestling matches. These occasions were marked by a great sociability. *Killing Time* itemizes these activities, but does not, I think, fully convey their atmosphere.

Take mealtimes, for example. Feyerabend would hold court at Berkeley's faculty lunchroom, in later years at the Chez Panisse restaurant. A steady stream of students, faculty, and assorted personages came and went. Paul was the main attraction. Discussions would careen through fine food and wine; music and opera; romantic prospects and dénouements; Feyerabend's revered classics of literature, history, and philosophy—perhaps he would be rereading an old favorite, perhaps a new book offered a novel treatment; Perry Mason, mystery books, and the best soap opera actors—soap opera being the sole repertory acting on TV; intellectual celebrity gossip; even philosophy of science. Feyerabend would discuss the state of his ailments, and his newest try for a remedy (e.g., acupuncture). Discussion would continue afterward, as Feyerabend with difficulty would make his way to a class or his bus stop; it was a big occasion when he got a specially outfitted car. His house was a dense labyrinth of books; indeed for years one of my functions was to scout out interesting books on any topic, and interesting musical discoveries, to recommend to him. These were immensely sunny times. Feyerabend was like champagne, sheer fun to be around.

Always there were jokes and gentle teasing; not infrequently there would be singing. You could tell that Feyerabend meant what he said in *Killing Time*:

> When at my best, I could do almost anything with my voice. I could let it go, rein it in, produce the softest pianissimo, and could increase my volume without feeling that I was reaching a limit. Singing gave me a great sense of power. My voice also projected well: soft or loud, it could be heard anywhere in a large concert hall. And it was beautiful, at least while I treated it well. For me, no intellectual achievement can give the joys of using an instrument of this kind. (pp. 82–83)

Aficionados of music and opera (I am one) will prize the many discussions of performances and personalities in *Killing Time*. The book is a Catalogue Aria of who slept with whom, who sang what, and always who was kind and humane, and who was rude or cruel. Feyerabend's voice also was a treat in class. This actor's undergraduate courses drew the largest enrollments on campus; once, only the university's Greek Theater had enough space. That suited Feyerabend to a tee.

He tells us that he was enthralled by the brassiness, the dazzling variety, the sheer un-Europeanness, of American culture, "a strange world that sustained a Mack Sennett, a Joe McCarthy, a Busby Berkeley, and a George Bush". America was "the first country that gave me a vague idea what a culture might be". Yet he often wanted to get away, often found Berkeley a cultural desert: "I am suffocating". His trips to London and New York were as oxygen to him. He would anticipate them and recount how they had picked up his spirits.

There are sections of *Killing Time* which intimate that Feyerabend's vivacity was at times a frenetic front, that he was a man deeply hurt, constantly fighting feelings of bitterness and aggrievement. Being crippled and cut off from life's wonders of children and sexuality would be bitter medicine for anyone, let alone someone of Feyerabend's personality and charm. I think Feyerabend's journey was in this respect a roller-coaster *Winterreise*.

Lovingkindness

> Papageno (*steht auf*): Aber sagt mir nur, meine Herren, warum muss ich denn alle diese Qualen und Schrecken empfinden?—Wenn mir ja die Götter eine Papagena bestimmten, warum denn mit so viel Gefahren sie erringen?
>
> Zweiter Priester: Diese neugierige Frage mag dein Vernunft dir beantwortem. Komm! Meine Pflicht heischt, dich weiterzuführen. (*Er gibt ihm den Schleier um*).
>
> Papageno: Bei so einer ewigen Wanderschaft möcht einem wohl die Liebe auf immer vergehen. . . .
>
> Papageno: Ein Mädchen oder Weibchen wünscht Papageno sich, O so ein sanftes Täubchen wär Seligkeit für mich.
>
> [Papageno (*stands up*): But tell me, Sirs, why must I undergo all these torments and horrors?—If the gods have truly assigned me a Papagena, why do I have to endure so much tribulation to win her?
>
> Second Priest: Let your reason answer this prying question. Come! My duty is to lead you onward. (*Covers Papageno's head with a veil*).
>
> Papageno: With such eternal wandering one might as well give up love forever. . . .
>
> Papageno: I'd give my finest feather To find a pretty wife Two turtle-doves together We'd share a happy life]

The capacity to give and receive love, to feel for others, is an ongoing theme in *Killing Time*. This capacity seems to have been more miss than hit throughout most of Feyerabend's life. He rather flatly records instances of cold or selfish behavior, of hurts to others along this path, and tries to reconstruct his 'mindset' at those times. For example he recalls a time when he "took evasive action" because "the more I was in love, the more I hated the slavery it seemed to imply". He came to grieve that he had pushed away his parents and neglected his father. Much of this interior life was embodied in Feyerabend's dreams, which are examined at decisive stages in his life.

Feyerabend says his quest succeeded when he and Grazia fell in love. "At long last I have learned what it means to love somebody . . . the long days with Grazia . . . turned me from an icy egoist into a friend, a companion, a husband". Feyerabend poignantly tells of his newfound emotional balance, his attempts to become a father—"After a life of fighting for solitude, I would like to live as part of a family . . . "—and above all of his hopes of growing old with Grazia.

He advises that fortune in love comes as an act of grace:

> Today it seems to me that love and friendship play a central role and that without
> them even the noblest achievements and the most fundamental principles remain
> pale, empty, and dangerous. . . . Love lures people out of their limited "individu-
> ality", it expands horizons, and it changes every object in their way. Yet there is
> no merit in this kind of love. It is subjected neither to the intellect nor to the
> will; it is the result of a fortunate constellation of circumstances. It is a gift, not
> an achievement. (p. 173)

So does goodness:

> Looking back . . . I conclude that a moral character cannot be created by argu-
> ment, "education," or an act of will. It cannot be created by any kind of planned
> action, whether scientific, political, moral, or religious. Like true love, it is a gift,
> not an achievement. It depends on accidents such as parental affection, some
> kind of stability, friendship, and—following therefrom—on a delicate balance
> between self-confidence and a concern for others. (p. 174)

I am glad that Paul received and gave these gifts in far greater measure than
he realized. Paralyzed by a brain tumor, he thirsted "not [for] intellectual survival
but the survival of love".

Conclusion

Is *Killing Time* true to its lead character? Does it do justice to the ideas and
personality of this marvelous man? I think so, for two reasons. The first is that it
sounds like Feyerabend: I cannot help but imagine the trademark Feyerabend
gestures and inflections that would have accompanied his printed words. The
second is that it *reads* like Feyerabend. This does not so much refer to his famous
books, but to my old letters and postcards from Paul, from the early 1970s to the
late 1980s. They reprise many things that appear in *Killing Time*, often in similar
phrasing: his jugglings of job offers; his operatic, scholarly, and literary enamor-
ments; and his generous advice on work, life, and love. If only I could write back.
There is so much to talk about.

Anakreons Grab
(Text: Goethe, Music: Wolf)

Wo die Rose hier blüht,
Wo Reben um Lorbeer sich schlingen,
Wo das Turtelchen lockt,
Wo sich das Grillchen ergötzt,—
Welch ein Grab ist hier,
Das alle Götter mit Leben schön bepflanzt und geziert?
Es ist Anakreons Ruh.
Frühling, Sommer, und Herbst genoss der glückliche Dichter;
Vor dem Winter hat ihn endlich der Hügel geschützt.

[Anacreon's Grave

With the roses in bloom,
with laurel and vines intertwining,
where the turtle doves woo,
here where the cricket is gay,
ah! whose grave is here?
That all the gods make alive with living things, and so fair?
'Tis here Anacreon lies.
Springtime summer and fall with joy he gave us his poems;
and at last now, in the winter is safely at rest.]

Notes

1. Quotations are meant to be sung with musical accompaniment, or read inwardly as if so sung. All translations are by the author, except for 'Anacreon's Grave', which is adapted by the author from the H. Drinker translation, and the final quotation from *Die Zauberflöte*, which is from the R. and T. Martin translation.

2. *Killing Time: The Autobiography of Paul Feyerabend*, (Chicago: University of Chicago Press, 1995). All quotations from Feyerabend are from this book.

3

Bas C. van Fraassen

Sola Experientia?

Feyerabend's Refutation of Classical Empiricism

Feyerabend's 'Classical Empiricism' (1970) draws on a seventeenth-century Jesuit argument against Protestant fundamentalism.[1] The argument is very general and applies to any simple foundationalist epistemology. Feyerabend uses it against Classical Empiricism—roughly, the view that what is to be believed is exactly what experience establishes, and no more—which he identifies as among other things Newton's "dogmatic ideology".

I will first examine the argument, then ask whether it is perhaps too powerful, a skeptical argument which, if cogent, undermines any pretense to rational belief. Feyerabend explains why the argument could have had little force in the targeted Protestant community. But that is significant only if we can think of its epistemic stance as rational. Feyerabend's irony leaves us with doubts, and yet, perhaps, also with some hope of a genuinely promising development.

1. The Jesuit Argument

The Jesuit characterizes his Protestant opponent as what might today be called a fundamentalist (I take no responsibility for this characterization): this position is characterized by the rule of faith: *sola scriptura*. This rule says that on religious matters, and indeed on anything on which the Scriptures pronounce at all, Scripture is the one and only source of information. Believe Scripture and Scripture alone—that is the rule. What if Scripture has something to say about a certain subject, but leaves some questions about it open? In that case the rule is presumably at least negative: to believe only, or at most, propositions in accord with Scripture.

In philosophical terms this is clearly a foundationalist epistemic position. The position identifies a basis or foundation for all rationally permissible opinion or belief and for all knowledge-claims in a certain domain. In epistemology, as in ethics, we should distinguish between positions and policies and ('meta-') views concerning them. "Foundationalism" sometimes denotes just the metaview that foundationalist positions (in the present sense) are the only viable epistemic positions.

The Jesuit argument has three parts. First, it is not self-evident what is and what is not (genuine) Scripture. Which texts belong to the canon? Of those texts that do, are any parts due to errors or additions made in transcription? According to the rule of faith, this question is to be answered in accordance with Scripture. In fact, Scripture *appears* to say a good deal about it, as when the text records that some speaker is a prophet or says of itself that it was recorded by an eyewitness. But we cannot draw on this without circularity. As long as the identity of Scripture is in question, we have—by the rule of faith—no basis for determining what Scripture says.

Secondly, the meaning of (putative) Scripture is not everywhere apodictically clear, hence requires interpretation. Saint Augustine interpreted Genesis allegorically; this may be seen as violating the rule of faith. But the mere fact that competing interpretations are offered creates a question to be answered. Again, the rule of faith appears to be applicable, since (putative) Scripture includes many directions to the reader, by itself classifying some parts as poetry or song, some as parables, some as history. But the very passages that on one reading are directions to read something as history, biology, or astronomy are on another reading examples of familiar narrative devices common to many forms of fiction and dramatic literature. We cannot draw on Scripture to settle its own interpretation.

Thirdly, in the attempt to settle whether some belief is in accordance with Scripture, we need to know how to draw out its implications. Pure logic will not get us anywhere. It is not a matter of logic that all humans have hearts, kidneys, knees, and elbows. Yet Scripture need not say explicitly that Abraham did, for us to take it that he did. In assimilating ordinary descriptions we draw consequences via general background and default assumptions of our own. However, we cannot interpret the rule of faith as allowing us to add all our own opinions wherever; as far as pure logic is concerned, Scripture is silent. (This is especially evident in the case of miracle stories). To draw on Scripture we must settle what it implies; but we cannot draw on Scripture to do so.

To summarize: we cannot apply the rule without identifying, interpreting, and extrapolating from Scripture. But each of these admits of alternatives. The rule would ask the impossible, namely that choice between these alternatives should be on the basis of Scripture itself. Thus the formulated epistemic position is untenable. Its problems are in fact logical problems, having very little to do with what Scripture is or is meant to be.

2. Is There Any Way Out?

Feyerabend extrapolates the Jesuit argument very quickly to apply to classical empiricism in which—he claims—experience plays the role of Scripture. We may doubt the analogy, given that experience and texts are prima facie very different.

But before we examine the asserted parallel, let us look at the Jesuit argument itself.

There appear to be two possible responses. The first is to say that the Jesuit's three problems do not arise at all, and need no solution. The second is to say that he is right, but we have an additional source of information. Obviously he himself opts for the latter, with (church) tradition as the needed arbiter.

How could it be that the three problems do not really arise? One could say: we do not need to try and identify Scripture because we already know what it is; we don't need to interpret it because its meaning is clear; and we have no difficulty deriving its consequences, which must be done by the same rules as for any other text. Feyerabend identifies this as the actual Protestant response, made possible because the rule *sola scriptura* is announced in a *community in which everyone already uses the word "Scripture" to refer to the same text*, and which already agrees in its reading of that text. But this means that there is in effect a second authoritative source of information within that community, namely, what it designates as "our" understanding—in other words, its tradition. The existence or even possibility of a differently constituted community prevents that tradition from having the force of pure logic. Therefore the first response reduces to the second.

But does the Jesuit argument not work equally well against any and every putative foundation for knowledge? If so, it should apply to the view that correct opinion is opinion based solely on Scripture and tradition, or on Scripture identified and interpreted according to one particular tradition. Surely the tradition cannot be used to answer questions like: what, precisely, is our tradition, what does it include and exclude, how is it to be understood? Or rather, it can so be used only with the same vicious circularity as was imputed above.

If that objection is cogent, then the Jesuit argument is more powerful than it looks, and destroys even what one might surmise to have been the Jesuit's own conclusion. But the argument still need not lead to skepticism. For its application is still solely to *foundationalist* positions. The proper conclusion is then: only non- or antifoundationalist positions can survive this critique.

Now we see where Feyerabend is headed, or think we do. Somehow the classical empiricist will be in the depicted Protestant's position with experience in the role of Scripture. The ensuing critique will be survivable only by a nonfoundationalist position. But can any position still allow for rational mutual criticism if the basis for such criticism (Scripture, experience) is deprived of its epistemic primacy?

3. Analogy: Experience/Scripture

Classical empiricism points to a unique putative source to ground and test all judgment: 'experience is our sole source of information'. A naive appeal to experience assumes that there is never any question about what the deliverances of experience actually are, nor about their meaning or significance. It assumes furthermore that the implications—namely, which theories are in accord with experience and which in conflict—are evident and unequivocal. But these assumptions are not tenable, as becomes clear as soon as we turn from the philosophers' idealization to actual practice.

There are three main problems, ignored by such a naive point of view, not coincidentally parallel to the Jesuit's problems for the rule of *sola scriptura*. The first is the identification of what is experience in the requisite sense. In one sense, any report we can give of what we think has happened to us is a report on our experience. But that is useless without certain distinctions. Suppose I come in from the garden and report seeing a yellow flower. Perhaps what I actually saw was a candy wrapper, or perhaps I saw only grass but had a small hallucination or fainting spell. In all these cases my report was still presumably veridical in the *minimal sense that indeed, it seemed to me* that I saw a yellow flower. However, as basis for knowledge claims, as evidence, or as test for opinion, that is useless. For such a basis, we would need to identify those experiences that are veridical in a further sense, namely, indicative of what it was that was actually seen, touched, or heard.

Secondly, even after that identification is made or assumed, there remains a dubitable element of interpretation. Two people who have looked into a furnace will report on, respectively, oxidation and phlogiston escape. The terms learned at their mothers' knee are theory-laden, so the report is infested with theory. Thirdly, even if identification and interpretation are held fixed, an almost century-long effort to codify evidential relations (so-called "confirmation theory") should have convinced us that "in accord with experience" is not a simple, uncritically usable notion.

So far so good for Feyerabend. But the disanalogies are not to be ignored. The analogy looks fine if we simply use the noun "experience"—a noun like "scripture", "source", "text". But this noun is a philosophical prop that we cannot afford to use naively. One major historical (empiricist?) confusion conflates experience in the sense of *events which happen to us,* and of which we are aware, with the *judgments* involved in this awareness. For example: I stepped on a garden hose, but I took it to be a snake—I jumped and screamed, so everyone noticed my mistake and laughed at me. The event that happened to me, and of which I was certainly aware, differs from the content of my response: the judgment that I was stepping on a snake. The two are by no means the same, nor inseparable, whether conceptually or really. This is clearest when the judgment is mistaken, but equally correct when it is true. This does not mean that one of these is the experience and the other something else; but we shall be confused if we either conflate the two or ignore one.

With this distinction in mind, how do the Jesuit arguments transpose? I have no problem of identification for the events that happen to me—at least, not in the sense that I must find some criterion to isolate the events that happen to me among all the events that happen. On the other hand, I have no problem of interpretation for the judgment I make in response. For any such spontaneous judgment, if explicit, is made in my own language. But this judgment does identify—in the sense of *classify*—the event in question. It also involves an element of interpretation, because it is couched in my own language which even I myself, on reflection, recognize to be heavily laden with old beliefs and theories. Worse, the text comes shrouded in uncertainty, ambiguity, and inconsistency.

So what does the rule of *sola experientia* prescribe? What I have experienced, in the sense of what has really happened to me, is the touchstone for all theory. But theory is, in me, confronted only with the text of my spontaneous judgments, that is, my immediate judgmental response to what befalls me. This text must be divided into dreaming and waking; and it must also be subjected to a critique in which I isolate at least a first layer of interpretation in the text. What shall I take as guide for these tasks?

In practice I will certainly trust and rely on my prior opinion and theoretical commitments. I will police my own data, and not accept any immediate, spontaneous, unreflective responses as ultimate authority. This is the analogue of relying on tradition. But doing so, I can no longer pretend to empiricist foundationalism: *sola experientia* was not a rule I could follow strictly at all. Foundationalists will now tell you that at this point I will inevitably slip into a debilitating skepticism or relativism.

4. How Could the Jesuit Escape Skepticism?

The reason that the 'tradition' response looks cogent at first sight may indeed be because it can be taken as an attempt at a nonfoundationalist epistemology. The idea is this: the rule of faith itself is a (prescriptive) statement, and has no meaning at all if regarded simply as a bit of syntax. Like any statement, it is meaningful only if heard as a statement in *our* language, with both sense and reference, therefore, fixed *for us* to some reasonable (or sufficient) extent. This is equally true of the questions that can be raised about it. We cannot think of our understanding or interpretation as a further text belonging to a language with less semantic structure than our own, which we could then believe or disbelieve, doubt or dispute.

In this way the objection is countered: the Jesuit argument applies to any foundation that can be represented as a text in a language in which the sense and reference of that text are not already fully determined. Relying on tradition does not mean believing in an additional text that informs us of the meaning of all other texts, but consists rather in the indispensable reliance on our understanding of our own language. That reliance is a precondition not only of the meaningfulness of the rule of faith, but equally of any discussion thereof. Of course, our tradition, in this sense, has no power of legislation against other identifications and interpretations, offered from outside—except for *us*, whose tradition it is and who persevere in our adherence to that tradition!

In some respects this response is clearly rather feeble. As a Protestant reply to the Jesuit, it would seemingly give up on the use of Scripture as an incontrovertible arbiter between them. Maintaining the rule of faith as understood would admittedly amount precisely to self-assertion, a matter of will, namely determined adherence to one's *own* tradition. As a Catholic alternative to Protestantism it would amount, after explicit addition of the texts codifying their tradition, to an exactly similar stance—namely, determined maintenance of their own form of understanding of Scripture.

There is a possibility for both to make their differences explicit, and attempt to engage in a dialogue "on common basis", that is, with those differences suspended. How optimistic can we be about such a dialogue? The usual dilemmas of (even moderate) relativism appear: how to respect the coexistence and right to life of alternative beliefs and attitudes without giving up one's own?

Yet it is clear that we have not necessarily arrived at a debilitating or self-destructive form of relativism or skepticism here. Rather, from both an intellectual and (in a broad sense) political point of view, this is our actual situation. The philosophical problem is not to find the sort of foundation that would rescue us from this aspect of the human condition. It is rather to account clearly for how we can live and function epistemically perfectly well (as we sometimes do!) under these conditions.

To summarize then: the Jesuit argument does not lead to skepticism but only to a rejection of any position that posits a foundation representable as a text. For we cannot draw on a text in any way without relying on something else, if only on our own language. This is true equally whether we regard the text as being in our own language or as translated into our language. But what we rely on is not itself representable as a text or body of information, so the same questions do not arise. On the other hand, it clearly admits of alternatives, and so this way out does not allow for Scripture—or any other source—the sort of role that foundationalists wanted a foundation to play.

5. The Uses of an Unfollowable Rule

Let us put the issue of foundationalism behind us once and for all. We can't be foundationalists, least of all in epistemology. We have to accept that, like Neurath's mariner at sea, we are historically situated, relying on our preunderstanding, our own language, and our prior opinion *as they are now* and go on from there. Rationality will consist not in having a specially good starting point, but in how well we criticize, amend, and update our given condition.

Fine, but now the Protestant and the empiricist still come along with their rules of faith. Could we see some meaning in them that makes them relevant in our postfoundationalist condition?

Feyerabend correctly points out that the *sola scriptura* rule plays two very effective roles in the community when it holds sway. First of all, because there is a tradition that identifies and interprets Scripture, the rule tells everyone who lives in that tradition not to depart from it. So it reinforces and maintains orthodoxy. He correctly likens it to Newton's famous Fourth Rule of Reasoning in Philosophy—let me state it here in today's idiom:

> In experimental philosophy we are to look upon propositions collected by general induction from phenomena as accurately or very nearly true, not withstanding any contrary hypotheses that may be imagined, till such time as other phenomena occur, by which they may either be made more accurate, or liable to exceptions.

This tells us that no new or nontraditional interpretations of the source material are to be heeded—only (other) source material can be consulted as touchstone for our opinions.

Is this a rationally compelling rule? Surely not. It is a counsel of epistemic conservatism, for which we may see some good practical reason. But prudence comes in degrees, we all have our personal risk quotient, and it is hard to see how logic could bend us one way or another.

But besides this obvious first role, the rule has a second equally effective role in the critique of accepted opinion. Should the need to revise our understanding arrive, the means are at hand. For suppose we want to change our belief from A to B. There is exactly one acceptable form of argument. We admit first of all that A was indeed solidly based on Scripture (or on experimental results, as the case may be). But then we point out interpretative elements in A, extrapolation or generalization not *logically* implied by that source. Suspending that part of A we are left with a weaker proposition A*, compatible with B. Now we must, as the third step, point to a part of Scripture (or experimental results) that together with A* provides support for B.

This is clearly a format that relies heavily on a shared preunderstanding, while making it possible to effect an agreed change in that understanding. Note the illusion traded on, namely that the correct opinion is uniquely determined by the source. Given this illusion we can point to the interpretative elements in A as a mistake, an 'un-scriptural' or 'unscientific' addition. This is how popular science and school texts now treat Newton: he did not arrive at Einstein's relativity because, enslaved by old ideas and seeing his own results with myopic eyes, he extrapolated his facts in a biased way. Einstein pointed out how Newton had gone beyond the deliverance of experience, removed Newton's metaphysical additions, and thus made way for the right theory, truly true to experience.

The form of argument I described was for a way of 'talking oneself into' a new view at odds with the old. Disturbingly we see here (and Feyerabend trades on this) the possibility of justifying any such change in view in this way *in retrospect*. Earlier experience cannot logically imply what later experience will be like. Therefore if some opinion based on earlier experience does not fit the new, it must have gone logically beyond what was given–of course! By its failure it stands convicted of overinterpretation, for experience (or Scripture!) itself could not have been the source of error! The pattern Feyerabend elicits here is familiar and convincing. But that the argument form has this use for retrospective rationalization does not mean that it does not also have actual use as a format in which the community can cooperatively talk itself into a new view.

So in its second role, the rule prescribes the acceptable form of critique. Can *that* role be justified? Not a priori. It expresses an attitude, approach, or stance taken by this particular community. In addition, it too admits of degrees as to how much it takes for a critique to become sufficiently weighty to carry the day.

In its two main roles, the rule of *sola scriptura* or *sola experientia* is apparently at war with itself. In one role it maintains orthodoxy and forbids heeding the alternative interpretations ingenious minds can concoct. But in its other role, it devalues any aspect of orthodoxy that can be identified as interpretative—and thus licenses a procedure that can successively peel off layer after layer. In the hands of logicians run wild, it would destroy the shared understanding altogether. But in a stable intellectual community, logicians are not allowed to run wild.

We are here face to face with an aspect of the social dimension of reason that philosophers are not happy to grant significance. The communal opinion remains stable because it is only philosophers who take ideas to their logical extreme, and happily no one listens to philosophers. In the hands of "reasonable" people, this dual-role rule is a great boon. It will maintain the status quo as long as there are no serious anomalies or new situations that throw the tradition into crisis. When such crises do happen, that same rule (in its other role) shows the way to reasoned and proportionate change, providing a rational form for consensual revision. The philosopher, at least if of rationalist or analytic bent, will be taken aback to see that reason, within a community, becomes a matter of politics, broadly speaking, and not of logic. *Reasonable* is not an epistemic but a bourgeois virtue.

There is a disturbing insinuation of double-think, bad faith, and self-deception within Feyerabend's praise for this Janus-faced rule of faith. The mariners-at-sea image depicts our real situation, but can it sustain itself only through the illusion of foundations? Is authenticity not possible? Are we dealing here merely with the last defence of a waning orthodoxy? Or can we reframe this practice so that it becomes at once a reasonable reliance on what we trust ourselves to have gained so far *and also* allowing of a sufficiently detached critical stance toward our past?

Feyerabend discusses this here only briefly, cryptically, ironically. If the rule of faith is actually empty but available for polemical use, we gain a terrible new freedom. Social factors, party lines, will drive debate when rationally compelling grounds are missing. But:

> Party lines are not the problem. Problems arise only when an attempt is made to turn the *subjective* conviction that makes a certain party line stand out into an infallible *objective* judge who withstands criticism. . . . [T]he democratic way in which praise, blame, and dogmatism are now distributed and the humanitarian way in which the word of a clever man is taken seriously, even *too* seriously, allow us to greet [this 'Protestant'/'empiricist' practice] as the dawn of an even more enlightened future. (Feyerabend (1970) 1981, p. 51)

This is still Feyerabend at least halfway into ironic mode, but there is a serious conviction at the core of his irony. In his own work, going forward from this critique of bygone epistemologies, we see him struggling with this task. The task not yet understood in the seventeenth century is as yet unfinished in the twentieth. Feyerabend put his shoulder to the wheel with the rest of us, even if he liked to chastise us scathingly from time to time for our laggard ways.

6. Concluding Comment

Feyerabend's argument does refute the position he characterized as classical empiricism. Whether or not that is an actual historical position, the argument is important because it affects a large range of conceivable positions. As we have seen, there are ways in which its targets might escape its force or amend themselves so as to become tenable. Since Feyerabend often enough called himself an empiricist, he might have been in sympathy with this attempt to see what empiricism could become, in view of his demonstration of what it could not (tenably) be.

I am willing to follow Feyerabend some way in this view of our epistemic life. In the normal course of things we merely update our opinion logically (nonampliatively) in response to experience. A crucial role in this process of spinning out the implications of our prior opinion is that of 'expert sources'. These are the explicit occupants of the role of 'tradition', but they are expert for us by virtue of our so regarding them. (This attitude toward science, tradition, newspapers, statistics gathered for certain reference classes, etc., is also a type of epistemic attitude, along with the 'lower level' ones of simply factual opinions and beliefs.) But we enjoy a certain freedom that allows us to take up, at any moment, a more detached critical attitude toward our epistemic life so far—and, given a certain accumulation of disappointments, to proclaim it a failure in certain respects, ready for pruning, slashing, or burning. There is however no rule or recipe, and there can be no such rule, to dictate how we must do this. We can demarcate certain forms of prudence, and insist on coherence, but in the end we face, in such 'revolutionary' moments, radical and unmediated responsibility for our own epistemic future. Feyerabend has pointed us here to notions of rules 'governing' (in some weaker sense than univocal prescriptive dictates) the negotiations and practices with which—perhaps?—we steer ourselves through such critical episodes.[2]

Notes

1. Feyerabend draws on the account of Popkin (1979), pp. 70–82. I shall discuss Feyerabend's construal of the Jesuits' (François Veron's) argument, with no presumption that it is historically accurate, nor even that it matches Popkin's understanding of the same material. My interest is in how Feyerabend's argument bears on the possibility of empiricist positions in epistemology.

2. The author wishes to thank Elisabeth Lloyd, Philip Kitcher, and Alvin Plantinga for helpful discussions.

References

P. K. Feyerabend (1970) 'Classical Empiricism', in R. E. Butts and J. W. Davis (eds.), *The Methodological Heritage of Newton* (Oxford: Blackwell), pp. 150–166. (Reprinted in P. K. Feyerabend (1981), *Problems of Empiricism: Philosophical Papers*, Vol. 2 (Cambridge: Cambridge University Press, 1981), pp. 34–51).

R. H. Popkin (1979) *The History of Scepticism from Erasmus to Spinoza* (Berkeley: University of California Press).

4

Peter Achinstein

Proliferation

Is It a Good Thing?

"So, I exaggerate"

— Paul Feyerabend

I served on several panel discussions with Paul Feyerabend. The most memorable occurred in 1970 at the University of Cincinnati. Paul delivered one of his spellbinding lectures in which he talked about Galileo's tower argument and claimed that Galileo was violating standard rules of logic and methodology. He was practicing "counterinduction" (which, according to Feyerabend, is a good thing to do). He was introducing a hypothesis, namely, that a body falling from a tower has a slanted, not a straight, motion, which is incompatible with highly confirmed theories and indeed is refuted by what we observe when we see the body fall. Galileo was following Feyerabend's favorite principle:

> *Proliferation I*: Invent and elaborate theories that are inconsistent with the accepted point of view, even if the latter should happen to be highly confirmed and generally accepted.[1]

In my response I tried to go through Galileo's argument, with numerous textual references, arguing that Galileo was not doing at all what Feyerabend claimed. Feyerabend's reply to me began as follows: "So", he said with a twinkle in his eye, in his charming accent, "I exaggerate".

Feyerabend exaggerated a lot. He espoused, or seemed to, extreme doctrines, the most famous of which he called "anarchism": there are no universal rules of reasoning in science. Any rules of this sort, rules such as the four that Newton proposed at the beginning of Book 3 of the *Principia*, are, and ought to be, frequently violated. Exaggeration, however, suggests that there is a truth there some place, even if it has been distorted or magnified beyond the fact. Is there a truth to proliferation? Perhaps there is, but matters need sorting out.

To begin with, if we understand proliferation in accordance with Feyerabend's explicit formulation above, there is nothing particularly shocking about it. Indeed, scientists and others practice it all the time. Suppose that I want to argue the advantages of a certain theory I favor that explains the facts in some way. Part of my doing so might involve my inventing a different, conflicting theory that also explains these facts but, I demonstrate, is incompatible with other principles that are 'highly confirmed and generally accepted' whereas my theory is not. A favorite theory is frequently defended by "inventing and elaborating" alternative theories in such a way as to satisfy Feyerabend's principle of proliferation. I take this to be straightforward and fairly trivial. Since Feyerabend wants to shock and exaggerate, I doubt that this is what he has in mind. The principle of proliferation, so understood, has no teeth.

Let's give it some teeth. One way to do so is to strengthen Feyerabend's injunction that we are to "invent and elaborate" conflicting theories. Scientists, we might say, don't just aim at "inventing and elaborating" theories for the sheer joy of it. They want to discover true theories, or ones that are probable, or ones that yield reasonably good predictions, or at least theories that are good in some important way (e.g., they are unifying or simple). Accordingly, one might strengthen Proliferation I to say this:

> *Proliferation II*: Believe or accept an "invented and elaborated" theory as true, or probable, or as a good predictor, or as good in some way, if it is inconsistent with the accepted point of view, even if the latter should happen to be highly confirmed.

This version of proliferation has a real bite to it. But it is absurd. First, many such theories can be invented and elaborated. Shall we believe or accept all of them as true, or probable, or good predictors, or good in some other way? That won't do, at least for truth and probability, if such theories are also incompatible with each other. More importantly, why should we regard a theory as true, or probable, or good in some way simply because it is inconsistent with the accepted point of view? I can think of no legitimate reason for doing so.

In any case, to my knowledge Feyerabend never asserts a principle as strong as Proliferation II. So what is he after? To determine this we need to identify his primary interest in formulating the principle of proliferation. What is such a principle supposed to do for us? On several occasions Feyerabend speaks of this principle in the context of the "growth of knowledge" and of obtaining "objective knowledge" (MS, pp. 22, 24; AM, p. 46). He also speaks of it in the context of "testing a theory" (MS, p. 26), and of giving us "evidence that might refute a theory" and thus detect error (AM, pp. 29, 41).

Perhaps Feyerabend is committed just to the following thesis, which I take to be more interesting than I and II:

> *Proliferation III*: Suppose we wish to test a theory T by determining whether there is evidence for or against it (so that we may obtain "objective knowledge"). The only way to test T—the only way to obtain evidence for or against it—is to invent and elaborate a conflicting theory T', even if T' is inconsistent with an accepted and highly confirmed point of view.

How is the invention of such a conflicting theory T' supposed to test theory T? One way noted by Feyerabend is that the invention of T' may enable us to unearth new evidence to test T (MS, p. 26). The theory T' may encompass phenomena that ought to be covered by T but were not thought of or appealed to in defense of T. Discovering such phenomena may then refute T. Unless an alternative such as T' were conceived, scientists would never have thought of the possibility of such refuting phenomena.

There is no denying this is possible. But this is pretty weak fare. If you have a theory T, then simply imagining a conflicting theory T' (where there are no constraints on T' other than perhaps that T' encompass some phenomena not appealed to in defense of T) does not guarantee or even make it likely that you will discover new evidence to test T. My theory is that O. J. Simpson committed the crimes for which he was charged. How is my inventing the conflicting theory that President Clinton did it going to help me unearth new evidence to test my O. J. theory? (Should I ask special prosecutor Starr to add this to his numerous investigations of Clinton by examining Clinton's logs for the day of the murders?). Simply inventing a contrary theory, without putting more conditions on it than Feyerabend appears to want, isn't likely to do much. I am much more likely to unearth new evidence by continuing to try to identify O. J.'s blood, his motives, his opportunities, his friends and neighbors, and so forth. Without more constraints on the conflicting theory T'—that it have some plausibility or some evidence in its favor—the possibility that it will enable us to unearth new evidence to test T is mere *logical* possibility. Moreover, it is clearly not the only logically possible way to do so. Concentrating on the evidence we have and asking more questions about it is another logical possibility, and it need not involve the invention of incompatible theories.

There is one situation in which the invention of a competing theory T' (without constraints on its plausibility) is important in testing T. Suppose you defend your theory T hypothetico-deductively by showing that from it a range of facts follows, some of which are known to be true, and others of which are predictions that are later verified. As Isaac Newton and John Stuart Mill pointed out a long time ago, this "consequentialist" test of T is not sufficient by itself to establish truth or even significant probability, since some competing theory T' may give the same results. Finding such a competitor, then, will serve as a further, and crucial, test of T. If the only argument offered in favor of T is that it generates a set of known and predicted facts, then "inventing and elaborating" a competitor that does equally well is an argument or test that at least blunts the force of the consequentialist argument.

It does not follow, however, that the invention of a competing theory T' which yields all the facts that T does (and perhaps more) will always test T in this way. This is a point emphasized by Newton in Rule 4 of his *Rules of Reasoning in Philosophy*:

> *Rule 4*: In experimental philosophy we are to look upon propositions inferred by general induction from phenomena as accurately or very nearly true, notwithstanding any contrary hypothesis that may be imagined, till such time as other phenomena occur, by which they may either be made more accurate, or liable to exceptions.[2]

Rule 4 needs to be considered together with the other three. Rule 1 enjoins us to admit no more causes than are true and sufficient to explain the phenomena. Rule 2 insists that we assign the same causes to the same effects. Rule 3 requires us to generalize inductively from properties present in all observed bodies to the presence of those properties in all bodies whatever.[3]

Newton uses his four rules in proving the law of gravity. The planets and their satellites exhibit the same observed effects: they are drawn off from rectilinear motion by forces that vary inversely as the square of the distance of the planets from the sun and the satellites from their planets. By Rule 2, we can infer the same cause or causes in all these cases. By Rule 1, only one cause is needed here, namely, the same force of gravity operating in the case of all the planets and their satellites. By Rule 3, we can generalize this from the observed motions of the planets and satellites to all bodies in the universe. As Newton does in Proposition 7 of Book 3, we can infer "that there is a power of gravity pertaining to all bodies, proportional to the several quantities of matter which they contain".

Now the point of Rule 4 is to say that we can make such an inference and regard it as true (or very nearly so), even if we can imagine some contrary hypothesis that explains the phenomena. If Rules 1–3 are employed to generate a universal proposition such as the law of gravity, then, Newton is claiming, the fact that some contrary hypothesis has been "invented and elaborated" from which known phenomena can be derived does not at all weaken the argument in favor of the universal proposition. So, for instance, suppose you can imagine that there is no universal force of gravity but a different force for each pair of bodies that has imperceptively different effects in each case; or suppose you can imagine a grue-type universal force that is an inverse-square force until the year 2500 but not thereafter. Such imaginings will in no way test or diminish the effectiveness of the argument in favor of universal gravitation, if that argument is constructed so as to satisfy the first three rules of reasoning. That is Newton's claim.

Let me generalize Newton's claim. Suppose that from what is observed (the "phenomena"), using some acceptable form of reasoning (whether Newton's rules or some others), one argues that some proposition is true or very probable. The fact that you can imagine some contrary proposition that explains the observed facts casts no doubt whatever on the original conclusion. This is a generalization of Newton's claim to which, I believe, he is committed. It is also a claim that Feyerabend seems to deny. How could he? I see at least three possibilities.

1. *Imagining a contrary hypothesis will enable us to challenge the proposition inferred by providing the means to challenge the "observed facts" upon which the inferred proposition is based.* It will enable us to see that some of these "facts" are not really facts. This is what Feyerabend claims about Galileo's tower argument. Galileo imagined a set of hypotheses contrary to the stationary earth hypothesis, namely, that the earth turns on its axis and that a body falling from a tower has two motions, one toward the center of the earth and one in the direction of the earth's motion, which together result in the body's having a slanting motion. These contrary hypotheses challenge not just the original hypothesis (that the earth is stationary), but the observed fact that forms the basis for this hypothesis, namely, that we see the body falling from the top of the tower not in a slanting line but a line that "goes along [the tower] grazing it, without deviating a hairs-breadth to one side or the other" and landing at its base.[4] Galileo's contrary

hypotheses enable us to see things differently and so reject the "phenomena" that serve as a basis for the original hypothesis.

I have two replies. The first is similar to one noted earlier. Simply inventing a contrary theory, without putting constraints on it, isn't guaranteed or even likely to show us that something we took to be an observed fact is not so. This is simply a *logical* possibility, no more. Suppose Galileo had imagined a Cartesian demon that makes it appear that the body is falling straight down; or that a body falls with a slanting motion but when it does so light is refracted so as to produce the appearance of a straight motion; or that there are twenty different motions of the body, resulting in some quite different curve we cannot see. Would, or should, these imaginings show us that what we took to be the observed facts are not really so?

My second point pertains to what Galileo was in fact doing when he proposed his contrary hypotheses. He was not claiming that we don't see the body fall "grazing the tower without deviating". We do see this. Nor was he claiming that the contrary hypotheses he was imagining entail or suggest that we don't see this. What Galileo was proposing was a conflicting explanation for this observed phenomenon, namely, that when we see the body falling what we see is the relative motion of the body with respect to the tower, not its absolute motion (which is slanted). Galileo is not challenging the observed facts but only the standard theory used to explain them.[5]

2. *Imagining a contrary hypothesis will enable us to challenge the mode of reasoning employed to defend the original hypothesis.* As noted earlier, this is the case when a hypothesis is defended hypothetico-deductively solely by reference to consequences. Imagining a contrary hypothesis that yields the same conclusion does or should serve to challenge the mode of reasoning (as Newton and Mill observed). But this is because the mode of reasoning is faulty and deserves to be challenged in just this way. If you claim that your hypothesis is true or probable solely because it entails a range of phenomena that are observed, I can challenge your claim by producing a contrary hypothesis that does the same thing.

Suppose, however, we consider Newton's Rules of Reasoning (at least Rules 1–3), and use them, as Newton did, to infer his universal law of gravity. Does imagining a contrary law not inferred from the phenomena by using these rules cast doubt on the rules themselves? Let's consider two cases. In the first, using Rule 2, from similar inverse-square effects of the planets and their satellites—that they are continually drawn off from rectilinear motions and retained in their orbits—and from the nature of these orbits, Newton infers that the cause that produces these effects in each case is an inverse-square force; and, using Rule 1, he infers that this is the same force (gravity) in each case. To invent a contrary hypothesis, we might suppose that the forces operating between the planets and their satellites, the planets and the sun, objects falling to the earth and the earth itself, etc., are all different in some respects other than being inverse-square forces. Or we might suppose that the force in question is the same in all these cases, but that it is not one force, but several different forces acting together to produce the resultant gravitational force. Doing so would violate Newton's Rule 1, which urges us "to admit no more causes of natural things than such as are both true and sufficient to explain their appearances".

In the second case, using Rule 3, from the fact that the law of gravity is satisfied by all the observed planets and satellites Newton infers that it is satisfied by all bodies whatever. To invent a contrary hypothesis (taking a cue from contemporary quantum mechanics) we might suppose that the law of gravity is satisfied by bodies only when they are observed, and that some different law operates when they are not observed. Or we might suppose that the law is satisfied until the year 2500 but not thereafter.

Does imagining these contrary hypotheses cast doubt on the rules of inference themselves? Newton would emphatically deny this. To be sure, it might turn out that one of these contrary hypotheses, or some other, is true. That is a logical possibility. But even if it does, that would not be enough to impugn the validity of inferring the same cause from the same effects or generalizing from observed instances. Newton is not claiming that such reasoning is *guaranteed* to lead to truth. This point he makes explicit in Rule 4, which says that we may regard propositions inferred by induction from phenomena (and presumably also by causal reasoning, in conformity with Rules 1 and 2) "as accurately or very nearly true, notwithstanding any contrary hypotheses that may be imagined, till such time as other phenomena occur, by which they may either be made more accurate, or liable to exceptions". So Newton explicitly recognizes that the propositions he infers using his rules of inference may be inaccurate, liable to exceptions, and hence false as they stand. But what he is claiming is that this can be shown only by new phenomena, and not by imagining contrary hypotheses. Doing the latter only shows that the original propositions inferred using causal or inductive reasoning or both could turn out false. But that "could" is only a logical possibility and casts no doubt at all on the modes of reasoning (or the propositions inferred).

More generally, with any mode of nondemonstrative reasoning, it is logically possible for the premises in such reasoning to be true and the conclusion false. If this logical possibility is sufficient to cast doubt on the mode of reasoning, then every mode of nondemonstrative reasoning is suspect. This, of course, is Hume's skeptical claim. Is that also what Feyerabend is claiming? If so, one doesn't really need to "proliferate", to "invent and elaborate" contrary hypotheses, to make such a claim. Simply stating that the conclusion of a nondemonstrative argument could be false will do the trick. One need not bother saying how it could be false by imagining specific contraries.

Feyerabend claims that his viewpoint is different from skepticism (AM, p. 189). The skeptic, he says, regards every conclusion "as equally good, or equally bad, or desists from making such judgments altogether" (p. 189). Feyerabend, by contrast, calling himself an "epistemological anarchist", wants to be allowed "to defend the most trite, or the most outrageous statement":

> The one thing he [the epistemological anarchist] opposes positively and absolutely are universal standards, universal laws, universal ideas such as 'Truth', 'Reason', 'Justice', 'Love', and the behaviour they bring along, though he does not deny that it is often a good policy to act as if such laws (such standards, such ideas) existed, and as if he believed in them. (AM, p. 189)

So in the end Feyerabend's view about rules of reasoning, such as Newton's four rules, seems to be this. Although generally it is "good policy" to act in accordance with them, to act as if we believed in them, they should not be *universally* followed. There are important occasions on which they are, and should be, violated. Rules of reasoning such as Newton's are not universally valid. They are good "rules of thumb", useful "guides" (see *MS*, p. 19). But like other such rules they need to be ignored or flouted on certain occasions.

If this is Feyerabend's view, then an important distinction needs drawing. The distinction is between, on the one hand, flouting, or violating, or ignoring a rule, and on the other, knowing when it is supposed to be applied and when not. As an example of the latter, I might refuse to use inductive generalization when I have very few instances, or when they are all of the same narrow type, or when there are some negative instances or ones that do not fit the generalization very well. Or I might refuse to infer a single cause from similar effects, if I know from other cases that effects of that kind can be produced by quite different causes. In such cases I have not flouted, violated, or ignored Newton's Rules 1–3. Rule 2, for example, does not say "To the same natural effects we must *always* assign the same causes". It explicitly says that we must do so "as far as possible". As is the case with most rules, there are facts about application that are left unstated and that may be unstatable in a way that will cover all possible cases. I have a pretty good idea of how to apply Newton's rules to various cases, when and when not to make the inference from effects to causes or from observed instances to all instances. When I refuse to make such an inference I am not necessarily flouting, or violating, or ignoring the rule, but simply realizing that it is not applicable in the case in question.

What about cases where the rule is flouted, violated, or ignored? There is a fifty-five-mile-per-hour limit on the road I am taking, but a passenger in my car becomes violently ill, so I exceed the speed limit to rush her to the hospital. Here I violate or ignore the fifty-five-mile-per-hour rule in order possibly to save a life. In such a case, I violate a rule to produce some desired state of affairs that perhaps will satisfy some more important rule. There are occasions when the rule can be ignored and violated for a better cause.

Which, if either, of these claims about rules of reasoning is Feyerabend making? If he is making the first, then his position becomes quite plausible. You aren't necessarily violating or ignoring Newton's Rule 2 when you refuse to use it to infer a cause from an effect. You may simply be applying it, as Newton tells you, only "as far as possible", and it may *not* be possible to apply it in a particular case. Similarly, you are not necessarily violating or ignoring Newton's Rule 3 when you refuse to generalize from observed instances. The latter may be too few or too unvaried to permit the inference, or there may be unexplained instances that appear to violate the generalization. Sometimes Feyerabend makes it sound as if this is what he has in mind. For example,

> The limitation of all rules and standards is recognized by *naive anarchism*. A naive anarchist says (a) that both absolute rules and context dependent rules have their limits and infers (b) that all rules and standards are worthless and should be given up. Most reviewers regard me as a naive anarchist in this sense overlooking

the many passages where I show how certain procedures *aided* scientists in their research ... I agree with (a) but I do not agree with (b). I argue that all rules have their limits and that there is no comprehensive "rationality". I do not argue that we should proceed without rules and standards.[6]

Feyerabend also makes the following claim, which might give some support to this interpretation:

> We may start by pointing out that not a single theory ever agrees with all the facts in its domain. And the trouble is not created by rumors, or by the results of sloppy procedure. It is created by experiments and measurements of the highest precision and reliability. (*MS*, p. 36)

Without here attempting to challenge Feyerabend's claim that every theory has counterinstances or at least ones that do not agree with the theory, if he is right, then use of rules such as Newton's in the case of "positive" instances becomes limited, to say the least.

On the other hand, there are many passages in which in addition Feyerabend seems to want to make the second, bolder claim about rules of reasoning, namely, that there are occasions (and they are numerous) when these rules ought to be ignored or violated, presumably for some better cause, where the latter is not simply to deal with known counterinstances or other failures of the theory. Thus,

> [m]ore specifically the following can be shown: considering any rule, however "fundamental", there are always circumstances when it is advisable not only to ignore the rule, but to adopt its opposite. ... There are even circumstances—and they occur rather frequently—when argument loses its forward-looking aspect and becomes a hindrance to progress. (*MS*, p. 22)

If this view is to be taken seriously, then we must imagine a case of the following sort (unlike the first sort of case). Newton has observed a range of planetary systems in which the gravitational effects are similar. No other phenomena are known indicating a different cause in each case. Newton's Rules 2 and 3 are applicable, in the sense that these rules are supposed to apply to a case of just this sort. Nevertheless, it is permissible, and even a good thing, to ignore or violate these rules by proposing different causes for each observed effect, or perhaps no cause at all, just chance, or by refusing to generalize beyond the observed systems, or by generalizing to a contrary hypothesis. It is permissible to do so in appropriate circumstances for a "better cause". What "circumstances", what "better cause"?

Feyerabend's answer to the "better cause" question seems to be that doing so "is reasonable and absolutely necessary for the growth of knowledge" (*MS*, p. 22). But how will it promote knowledge? The only answer that I can find is that we might discover that what is inferred (a universal gravitational force) does not exist, and that something else is responsible for the effects. We might discover error. Again, there is an appeal to a "logical possibility", which neither Newton nor I would regard as a "better cause".

What is Feyerabend's answer to the question of "circumstances"? In his discussion of Newton's Rule 4 he claims that legitimate criticisms of Newton's "ray" theory of light (according to which white light consists of rays of different refrangibility) can be unearthed only by ignoring Rule 4:

[Such] criticism can be articulated only if we are allowed to view the success of Newton's theory in the light of "contrary hypotheses" [thus violating Rule 4]. If on the other hand we follow Rule 4 to the letter, then contrary hypotheses will not be used and the criticism cannot arise. (IS, p. 401)

A few pages later we find

Instead of judging theories by a never-examined, mystical, and stable entity, "experience", one should let them compete with each other in the very same manner in which party lines are competing in politics. The invention of "contrary hypotheses" is the first step towards such a competition, and never is their invention more necessary than when it seems that certain ideas have been confirmed beyond doubt and that matters have been settled once and for all. (IS, p. 405)

Finally, in *Against Method*, Feyerabend writes: "Hence it is advisable to let one's inclinations go against reason in any circumstances, for science may profit from it (AM, pp. 155–156; italics his). Accordingly, the answer to "under what circumstances" it is legitimate to ignore or violate rules such as Newton's seems to be "always", if error is to be found.

At one point Feyerabend proposes a principle in addition to proliferation, which he calls the "Principle of Tenacity".[7] It urges us "to select from a number of theories the one that promises to lead to the most fruitful results, and to stick to this one theory even if the actual difficulties it encounters are considerable". Perhaps Feyerabend has in mind that in selecting such a theory in accordance with tenacity one is following rules of reasoning, while in proposing contrary theories in accordance with proliferation one is violating those rules (and practicing "counterinduction"). And perhaps he wants to say that while some in the scientific community should attempt to use rules such as Newton's as useful guides (as Newton did in arguing for his law of gravity, thus obeying tenacity), others should ignore those rules by introducing contrary theories that violate those rules (proliferation). Again, the proposed justification would be that such a practice, and only this, can detect error and produce "objective knowledge".

When Feyerabend says that there are circumstances when any rule of reasoning ought to be ignored (and its "opposite" adopted), I think it is fair to interpret him to be making not just the first claim about rules of reasoning, but the second as well. So, recalling his self-confessed penchant for exaggeration, one might say that he exaggerates by turning claim 1 into claim 2. He turns the plausible claim that you need to know how and when to apply rules such as Newton's, which are not applicable in all situations, into the implausible one that there are numerous occasions (perhaps always) on which rules such as these, although applicable, should be violated or ignored for a better cause.

3. *Imagining a contrary hypothesis will cast doubt on the hypothesis inferred even without casting doubt on the facts from which it is inferred, on the mode of inference, or on the claim that the hypothesis follows from the facts in accordance with the mode of reasoning.* So, for example, even if you accept Newton's six "phenomena" from which he infers his law of gravity (phenomena pertaining to the motions of the planets and their satellites), and even if you accept Newton's rules of reasoning (and the earlier propositions of Book 1 of the *Principia*) and

agree that using these the law of universal gravitation can be inferred, you cast doubt on that law simply by inventing a contrary hypothesis. You test, and ought to test, Newton's law in this way.

This is just what Newton rejects in his Rule 4, and I think he is right to do so. If you accept the mode of inference, and the claim that the conclusion follows, then how can imagining a contrary conclusion cast doubt? In such a case you cast doubt on the conclusion only if you challenge the facts (the "phenomena"). But, by hypothesis, these are unchallenged. So merely imagining a contrary hypothesis is (again) tantamount to pointing out the logical possibility that the conclusion could be false. And in science as in the law the mere logical possibility that a conclusion is false ought not to cast ("reasonable") doubt on a conclusion that otherwise has support. It is not a test of such a conclusion.

Notes

1. 'Against Method: Outline of an Anarchistic Theory of Knowledge', in M. Radner and S. Winokur (eds.), *Minnesota Studies in the Philosophy of Science*, Vol. 4 (Minneapolis: University of Minnesota Press, 1970), p. 26. There is a similar formulation in *Against Method* (London: New Left Books, 1975), p. 47. Hereafter I will use the abbreviations *MS* (Minnesota Studies) for the first work, and *AM* for the second.

2. Feyerabend discusses this rule in 'On the Improvement of the Sciences and the Arts, and the Possible Identity of the Two', in R. S. Cohen and M. W. Wartofsky (eds.), *Boston Studies in the Philosophy of Science*, Vol. 3 (Dordrecht: D. Reidel, 1968), pp. 387–415 (hereafter IS). I respond in 'Acute Proliferitis', pp. 416–424 of the same volume.

3. For a discussion of these rules, see my *Particles and Waves* (Oxford: Oxford University Press, 1991), chap. 2.

4. Galileo, *Dialogue Concerning the Two Chief World Systems* (Berkeley: University of California Press, 1967), p. 139.

5. In the *Dialogue Concerning the Two Chief World Systems*, Simplicio (the Aristotelian defender of the stationary earth theory) claims that "the senses . . . assure us that the tower is straight and perpendicular, and . . . show us that a falling stone goes along grazing it, without deviating a hairsbreadth to one side or the other, and strikes at the foot of the tower exactly under the place from which it is dropped" (p. 139). Salviati (the defender of the moving earth theory) responds by asking this question: "if it happened that the earth rotated, and consequently the tower, and *if the falling stone was seen to graze the side of the tower just the same*, what would its motion have to be?". Salviati's answer is that it would be a compound of two motions resulting in a "slanting". Hence, he concludes, "just from seeing the falling stone graze the tower, you could not say for sure that it described a straight and perpendicular line, unless you first assumed the earth to stand still" (pp. 139–140). Galileo is claiming that what we see is the same in both theories (we see the falling stone graze the tower and land at its base). The explanation each theory offers for what we see is quite different.

6. *Science in a Free Society* (London: New Left Books, 1978), p. 32. Italics his.

7. 'Consolations for the Specialist', in I. Lakatos and A. Musgrave (eds.), *Criticism and the Growth of Knowledge* (Cambridge: Cambridge University Press, 1970), p. 203.

5

John Watkins

Feyerabend among
Popperians, 1948–1978

D id I have a fitting piece for a volume in memory of Paul Feyerabend? Well, apart from a newspaper obituary,[1] I had just one piece devoted exclusively to him. Was it fitting? Not exactly; but a scene-setting introduction might help.

Paul Feyerabend and Karl Popper took a shine to each other when they first met, at the Austrian College in Alpbach, in 1948. Paul told me long afterward that they went for a walk, with Popper talking very easily, on a first-name basis, and he was impressed by the simplicity of Popper's arguments. However, in Vienna Feyerabend fell under Elizabeth Anscombe's influence; through her he briefly met Wittgenstein; and when he got a British Council Scholarship in 1951, he applied to study under him. But Wittgenstein died in April, and in 1952 Feyerabend came to LSE to work under Popper on problems relating to quantum mechanics.[2] It must have been then that he began to develop his fluent, racy style in both spoken and written English. But although I was teaching at LSE I don't remember meeting him then. (He afterward translated Popper's *The Open Society* into German.) I met him when he briefly passed through Alpbach en route to Berkeley in 1958, leaving a trail of broken hearts behind him. But it was not until 1961, when I and my family visited Berkeley, that I got to know him at all well.

The BBC has a radio panel game called 'Just a Minute' in which competitors are given some unexpected topic and required to speak on it for sixty seconds 'without hesitation, deviation, or repetition', starting now. Add that they are to speak with wit and inventiveness, and Paul would have been unbeatable. His inventiveness was challenged by our then five-year-old daughter's demand for a story; he came up with one off the cuff and without a moment's hesitation. She was thrilled, and thereafter was always besieging him with demands for stories. Eventually he became fed up and told her a story about a girl who was forever asking for stories; the news spread and people started arriving from all over the world with stories for her; and they insisted that she listen; if they arrived at night, they would come to her room and shake her, crying 'Story, story, wake up, wake up, you must listen to a story'. She never asked him again.

He read astonishingly widely (would any other philosopher's researches have led to the discovery that the Devil has an ice-cold member?). Sometimes he preferred the company of books to that of people over long stretches of time. (He told me of one period of several months during which he spoke to no one except the waiter in a small nearby restaurant where he ate.) And he was a great letter-writer. In 1973 I received an enquiry from the University of Sussex, as I was on the point of going on holiday, about his suitability for a chair there. I dashed off a short and truthful answer: he 'is one of the few people I know with real brilliance and originality. He is also wayward, erratic, moody and idiosyncratic. I personally find him a rather loveable person, but he can be pretty exasperating.' To introduce my above-mentioned piece on him I need to say something under this last head.

Once, after he had called on us at Berkeley one hot afternoon in an open-necked shirt, it turned chilly and I lent him my sweater. When we next met he was *wearing* it; he liked it and seemed to regard it as his. I said that I also liked it and regarded it as mine. (After some weeks a compromise was reached; he would return it and my wife would send him one just like it from London.) It was rather the same with ideas. He used to joke that when he visited our home in London he would first be conducted to the dressing-down room, as he called my study. In his autobiography (1995) he made it sound as though I would there chastise him for being insufficiently Popperian. But what I chiefly complained about was his being insufficiently forthcoming about how Popperian his ideas really were. In those days he regularly assailed what he called "empiricism", a position supposedly maintained by such thinkers as Ernest Nagel, Hempel, and Oppenheim and—or so it often seemed—Popper. But his attack used key ideas of Popper's.

This pattern was first exemplified in his 'Explanation, Reduction and Empiricism' (1962), of which he had sent me a stencilled preprint shortly before I came to Berkeley. I had some small influence on the published version; in a three-page 'Summary and Conclusion' there is a footnote acknowledging that Popper's (1957) had been the starting point for this essay. In Popper's *Objective Knowledge* (1972) there is an appendix, 'The Bucket and the Searchlight', whose content is superseded by chapter 5, a revised version of his (1957), 'The Aim of Science'. Then why include it? Well, a footnote says that it was given as a lecture at Alpbach in 1948, a coded way of saying that it had been listened to by Feyerabend; and a 'Bibliographical Note' attached to chapter 5 draws attention to the above-mentioned footnote in Feyerabend's essay (1962).

The two theses of Popper's 1948 Alpbach lecture that are relevant here are these. First, the primary task of science is *explanation*, which brings prediction and technological control in its train; and a necessary condition for premises to explain an explanandum is that the latter can be *deduced* from them. Second and more interestingly, when a new theory T_2, say Newton's, supersedes an earlier theory T_1, say Kepler's, what typically happens is that, far from entailing it, T_2 revises and hence and strictly *contradicts* T_1. (Popper used this as a decisive argument against the inductivist view of scientific progress).

In a footnote added to a reprinting of his (1962), Feyerabend pooh-poohed his indebtedness to this second thesis (theoretical progress is usually revisionary) which, as he rightly added, had in any case been stated already by Duhem.[3] But he told me that it had impressed him strongly when he first heard it in Alpbach. His attacks on "empiricism" mainly consisted of silently giving an additional twist to the first thesis (explanation is deductive) and then deploying this second thesis against it. He pointed out, very truly, that for an explanandum to be deducible from an explanans, (i) the latter must be consistent with the former, and (ii) those terms that occur in both must have the same meanings in both. He called (i) the *consistency* condition and (ii) the condition of *meaning invariance*. His twist was to add the assumption that if T_1 is a currently accepted and well confirmed theory, then "empiricism" requires the explanandum for any new theory T_2 that seeks to supersede T_1 to be T_1 *itself* rather than some revised version thereof. Given that addition, it does indeed follow that condition (i) will not allow T_2 to revise T_1 and condition (ii) will require ontological continuity between them; and that would of course have been a hopelessly conservative ideology for science.

But there is no conflict between those two theses of Popper's re-explanation being deductive and progress being revisionary. We have only to drop that addition and allow that T_2 may revise T_1, adding that if it does, then what T_2 explains is not T_1 as such but a revised explanandum T_1'. Kepler's laws said that the planets' orbits are perfect ellipses with the sun stationary at one focus. Newton's theory denied that, and explained instead why their nearly elliptical orbits are slightly distorted and why the sun is not quite stationary. In such a case, the new theory explains why its predecessor, though false, was nevertheless highly successful. These matters came up again in a long piece by Feyerabend, 'Problems of Empiricism' (1965). This appeared in a volume that I reviewed in (1966). I there said that this must be about the fifth time that he attacked the consistency and meaning invariance conditions, and I answered it briefly along the above lines. He wrote to me 'fifth time' should have been 'second time'; I now find that on the back of his letter I listed *six* places where he attacked those two conditions.

In the later 1960s Feyerabend was often in London. For a time it seemed that he might take up the HPS chair at University College London; and he gave several lectures at LSE. He and Lakatos took a shine to each other. (He got Lakatos invited to Berkeley in 1967; Berkeley is trembling with Imre-anticipation, he told me.) In December 1967 he wrote to me that Imre and I were very lucky fellows in comparison with him; we know what we want whereas he is now at a complete loss; he has been teaching Popperianism for over ten years; it is for him the most acceptable philosophy; and yet it now turns out to be a nightmare. My guess is that the nightmare factor had to do with his unwanted indebtedness to Popper. I formed the impression around this time that he had reached the rather desperate conclusion that his only way of escaping from this predicament and getting Popper off his back was to turn against critical rationalism and swing right over to some kind of irrationalism (he wasn't yet calling it anarchism or dadaism).

Being reluctant in that 1966 review to go on too much about debts to Popper I had concluded it with the remark: 'And a reference to J. S. Mill would have

been in place at the end, where he calls for an extension of his epistemological pluralism to extra-scientific walks of life.' He wrote to me that he had not
known about Mill; what book should he read? In January 1969 he wrote to
me that, having put Mill's *On Liberty* on the reading list, he had at last read it—
and found it marvelous; from now on it would be Mill to whom he would
refer. He was delighted to discover that whereas the illiberal and dictatorial Popper
had a wife, the liberal and humanist Mill had a *mistress* (Gellner was to comment
later that 'even the possession of a harem does not necessarily make a ruler liberal').

I will skip the 1969 "Troubles", that affected both Berkeley and LSE. By the
early 1970s Feyerabend and Lakatos, whose rift with Popper was now absolute,[4]
had become very close. Before turning to unhappier developments let me mention
an amusing episode from 1972. (One needs to know that Lakatos had injured his
neck and was wearing a surgical collar). A small conference in the history and
philosophy of science was to be held next summer in Jyväskylä, Finland. When
the organizing of it was nearly complete, it was discovered that I, who had not
been consulted, was officially a member of the organizing committee, and I was
belatedly brought in. I must have written to Paul saying that in my newfound
capacity I had tried to secure an invitation for him but gave up on being told that
the only remaining slot was in the history of Islamic science. He replied that he
was surprised by my ignorance of his work; with regards to Islamism he was the
foremost expert on one Lak-el-katos, the eleventh century rebel who was hanged
for his disrespectful treatment of Al-Poppuni, the reigning philosopher of the day
(and a great tyrant), but managed to escape with a bent neck after one of his
mistresses cut him down in time.

In late 1973 Lakatos, in collaboration with Spiro Latsis, was planning a conference on methodology in physics and economics, to be held next year in Nauplion,
a beautiful little port which, in the nineteenth century, had briefly been Greece's
capital. They lined up an impressive body of participants. Then, at the beginning
of February 1974, Lakatos suddenly died. I had not been involved with the conference, but when Latsis decided to carry on with it I offered to help. Everyone
wanted Feyerabend to take Lakatos's place. Lakatos's death had left us both in a
state of shock (he had been only fifty-one, and at the height of his powers) and I
had had a lot of communication with Feyerabend after it. I prevailed on him and
he agreed to give the opening talk.

July 1974 was, for me, a gloomy month. On July 3 I received a letter from
him enclosing a tape and saying that if, as seemed rather likely, he did not turn
up in person at Nauplion we could play this recording of his lecture instead. I
angrily returned it to him. On July 15 the "Colonels" then in power in Greece
carried out a military coup in Cyprus to overthrow Archbishop Makarios and bring
about enosis with Greece. On July 19 a Turkish invasion force set out for Cyprus.
War between Greece and Turkey seemed imminent and our conference doomed.
In the event, the Greek army commanders preferred toppling the regime in Athens to invading Turkey. The "Colonels" went, and democracy came to Greece.
So did Paul Feyerabend; my angry outburst worked. But our friendship never

recovered. He now cast me in the role of "the stern janitor of the Popperian temple'.[5]

Another consequence of Lakatos's death was that I became a coeditor, with John Worrall, of *The British Journal for the Philosophy of Science*. And an early problem was to find a reviewer for Feyerabend's *Against Method* (henceforth AM). I bumped into Ernest Gellner in the SCR at LSE one day, and he mentioned that he was reading it. When I asked him if he would like to review it for us he said he would. (Afterward Worrall told me that Lakatos had arranged for Noretta Koertge to review 'For and Against Method', as it would have been entitled if his as-yet-unwritten but intended reply to Feyerabend had been included; so I wrote apologetically to her explaining what had happened. She replied that she was glad that a "big name" was doing it). Gellner wrote a long review; it appeared in December 1975 and it was deadly. 'Character assassin assassinated' and 'Feyerabend out-Feyerabended' were typical comments.

Feyerabend poured out furious letters in various directions. With the help of my pocket-calculator I have estimated that during the next few months I received from him some fifteen thousand written words on or relating to the Gellner review. In one letter he said that having this review 'laid on me by you guys' was 'an extremely unfriendly act'. In public he wrote that when 'Watkins succeeded Lakatos as editor he replaced Koertge by Gellner';[6] if that means that I knowingly switched reviewing it from Koertge to Gellner it is not true. (I checked recently and found that it is by no means clear that Lakatos had actually said anything to her about doing it.) But I am inclined to plead guilty to an unfriendly act. My gut reaction to AM was that it is embarrassingly awful, and I knew, when I asked him to review it, that Gellner was steamed up about it.

My correspondence with Feyerabend spluttered on for about six months; near the end it veered round to Solzhenitsyn, who had recently been exiled from the Soviet Union on the publication of the first volume of *The Gulag Archipelago*. In those days I admired him greatly and eagerly seized on his books. Feyerabend had been watching him on television. What he thought of him can be gathered from my reply, which began: 'So: ponderous, depressing, puritanical creep Solzhenitsyn made fun-loving, anarchical Feyerabend want to puke . . . ' After some desultory exchanges in 1977 there came a five-year gap in our correspondence.

Feyerabend had declined an invitation to a conference in 1975 sponsored by the Fritz Thyssen Foundation and organized by Gerard Radnitzky and Gunnar Andersson, at which a group of us at LSE would present a "position paper". But he agreed to contribute to its published proceedings. He told me that his contribution would be a "punishment" for the above-mentioned unfriendly act. At around that time he also had a contribution in the proceedings of the Nauplion conference, and he made references to the LSE position paper there too. What follows is the section on Feyerabend in my essay, 'Corroboration and the Problem of Content-Comparison', (1978) in which I replied to criticisms. I have adjusted the references; it is otherwise unchanged. I may perhaps mention that the "letter to Santa Claus" referred to in it grew into the chapter "Optimum Aim for Science," set out and elucidated in my *Science and Scepticism* (1984).

Feyerabend's Effortless One-Upmanship

Feyerabend proceeds from a discussion of Aristotle to the conclusion that money should be taken away from critical rationalists. His method of "arguing" against us is to be always one-up by taking an assumption that was not under debate, declaring it to be "the question at issue", and then giving it his usual dadaist treatment. Our task in the position paper was to discuss criteria of scientific progress; it was not to discuss whether scientific progress is good or bad for mankind. This gave Feyerabend an easy opening: we dogmatically took it for granted that science is good and failed to examine the (meta-) question: *"what's so great about science?"* (1976a, p. 310). But suppose we had dealt with this question and that our answer had boiled down to this, that science seems to have done better than Azande magic, theology, etc., in its search for truth. Would that have met his complaint? Not at all. He would have complained that we dogmatically took it "for granted that Truth is something quite excellent" (loc. cit.) and failed to examine the (meta-meta-) question: "What's so great about truth?" (1978a, p. 167).

Feyerabend has said that his "favourite pastime is to confuse rationalists by inventing compelling reasons for unreasonable doctrines' (1975, p. 189). The chief unreasonable doctrine of his present essay is that we should discard logic. Let us see how compelling are the reasons he has invented for this. He claims that "there are facts whose only adequate description is inconsistent" (1978a, p. 165), but the moving-train example (p. 157) on which he bases this claim is so pathetic that I will not bore the reader with an exposure of it. He accuses critical rationalists of an "astounding dogmatism" in failing to use illogical theories as criticisms of the laws of logic (p. 155). Well, I once proposed that for a statement C to constitute a criticism of a theory T, C must have some sort of adverse implications for T and must itself at least be a candidate for rational acceptance, a minimal condition for the latter being that C is not self-contradictory (1971, p. 58). And I find no compelling reason here to change that. Feyerabend also says that "inconsistent theories are better to handle and lead to more discoveries than their decontaminated rivals" (p.165). What he means by 'better to handle' I do not know (nor, I suspect, does he); but I do understand the second part of this claim: the idea that contradictions are fertile and lead to progress seems to be the least uncompelling reason he has been able to find (he did not invent it, of course) for this unreasonable doctrine.

Before examining further what he says in this connection I wish first to quote a passage, written some forty years ago, from Popper's "What is Dialectic?" (1940). It will prove most constructive to compare this with what Feyerabend now says. Popper wrote:

> But the most important misunderstandings and muddles arise out of the loose way in which dialecticians speak about contradictions.
>
> They observe, correctly, that contradictions are of the greatest importance in the history of thought . . . without contradictions, without criticism, there would be no rational motive for changing our theories: there would be no intellectual progress.

Having thus correctly observed that contradictions . . . are extremely fertile, and indeed the moving forces of any progress of thought, dialecticians conclude—wrongly as we shall see—that there is no need to avoid these fertile contradictions. . . .

Such an assertion amounts to an attack upon the so-called 'law of contradiction'. (1963, p. 316)

Now read carefully the following sentences from Feyerabend's essay:

If we can give reasons for the usefulness of contradictions in science as I think we can then we have also reasons against the selected system [i.e. a system of logic that asserts noncontradiction]. . . .

Assume somebody shows as I think I have shown (with the help of . . . Hegel, Kuhn and Lakatos) that science often violates those laws of logic which our critical rationalists regard as a conditio sine qua non of rationality . . . Now if one discovers that knowledge is beset by contradictions, progresses because of the contradictions and *right through them* [my italics], . . . then the lesson is obvious: rationality as defined [by critical rationalists] has to be given up. . . .

But is it really so easy to get rid of fundamental principles of logic? (1978a, pp. 155–157)

The passage from Popper consists of categorical statements and its message is unequivocal: (a) contradictions cause progress because we have to work to eliminate them; (b) dialecticians wrongly conclude that contradictions can be retained because they cause progress. A hasty readings of the passages from Feyerabend might give the impression that they too contain an unequivocal message, namely that the fundamental principles of logic, including the law of contradiction, have to be given up. A close reading, however, reveals that *he is not actually asserting anything*. There is not one categorical statement in the passages I have quoted. He proceeds from a series of hypothetical statements ("If we can give reasons . . . ", "Assume somebody shows . . . ", "Now if one discovers . . . ") to an *insinuated* conclusion couched in the form of a question ("But is it really so easy . . . ?").

Suppose, however, that we replace those hypotheticals by categorical premises ("Contradictions in science *are* useful", "Science *does* often violate the laws of logic", "Knowledge *is* beset by contradictions, progresses because of the contradictions and right through them"), and the insinuated conclusion by the categorical conclusion: "Therefore the fundamental principles of logic have to be got rid of". How compelling or otherwise would this argument be? I say that any persuasiveness it might have for the unwary reader would be the result of its *fudging* of Popper's sharp distinction between claim (a) (contradictions cause progress because we have to work to eliminate them) and claim (b) (contradictions can be retained because they cause progress). The phrase where the fudging occurs is the one I have italicized: "Knowledge . . . progresses because of the contradictions *and right through them*", this is nicely ambiguous between (a) and (b). If it meant that knowledge progresses beyond the contradictions by surmounting them it would provide no reason at all for the present 'unreasonable doctrine'. If it meant that knowledge progresses because of contradictions that it *retains*, it would be exposed to the obvious retort: but if we do not have to work for the elimination of contradictions, how do they cause progress? Is not the doctrine that we can retain contradictions that have come to light a recipe for, among other things, intellectual laziness?

Now the difference between (a) and (b) is one of which Feyerabend must have been aware. (If he were not, he ought to be raising questions about *his* salary). For one thing, he knows that essay by Popper very well. (In the course of the passage from which I have quoted he refers to it (note 20): to p. 317; my quotation is from p. 316). The conclusion seems inescapable. He *wished* to confuse us by inventing a compelling reason for this unreasonable doctrine; unfortunately, this wish was not father to any new thought. So he decided to concoct a sham reason by deliberately obfuscating the clear and obvious distinction between (a) and (b); and to be on the safe side he incorporated this sham reason into a passage in which he actually refrains from venturing any categorical assertion. In this way the unwary reader would be given the impression that an argument for discarding logic has been provided, while the wary reader would be presented with no statement that is both unequivocal and false.

Feyerabend often complains that he is not read properly. I say that he often writes so that he *cannot* be read properly. Since that is a serious accusation I shall now digress, briefly, in order to support it with an independent example. Feyerabend has never made this complaint so frequently and bitterly as in his reply (1976b) to a scathing review of *Against Method* by Gellner (1975). In that review Feyerabend was quoted (on p. 340 with a reference to AM p. 187) as saying that violence is beneficial to the individual. To this he replied:

> [D]on't you think you should have read the text a little more carefully or let somebody else explain it to you in case you can't read? The text says that violence is necessary *according to political anarchism* and adds that political anarchism is a doctrine I reject. The very first sentence of the book calls political anarchism 'not the most attractive political philosophy' (p. 17) and on p. 189 I again distinguish my views from political anarchism, just to be on the safe side. All in vain. (1976b, p. 387, note 1)

That seems clear enough: Gellner has committed the reviewer's howler of attributing to an author a view that he had only reported, and moreover with disapproval. But let us investigate. To begin with, we find that on pp. 17 and 189 of AM he had been by no means as unequivocal as he here claims. On p. 17 he had written, not that anarchism is not the most attractive political philosophy but only: "*anarchism*, while perhaps not the most attractive *political* philosophy . . .". And on p. 189 he had given a characterization of 'the epistemological anarchist' who is there said to resemble "the Dadaist . . . much more than he resembles the political anarchist". There is in fact *no* clear-cut rejection of political anarchism in AM. And in any case the question of political anarchism is irrelevant: it is *not true* that "the text says that violence is necessary according to *political anarchism*". The paragraph from which Gellner quoted does indeed begin with two sentences on political anarchism: but political anarchism is then left behind and the passage continues:

> Occasionally one wishes to overcome not just some social circumstances but the entire physical world which is seen as being corrupt, unreal, transient and of no importance. This *religious or esc[h]atological* anarchism denies not only social laws, but moral, physical and perceptual laws as well and it envisages a mode of existence that is no longer tied to the body, its reactions and its needs. *Violence*, whether political or spiritual, plays an important role in almost all forms of anarchism. Violence is *necessary* to overcome the impediments erected by a well-organised society, or by one's own modes of behaviour (perception, thought, etc.), and it is *beneficial* for the individual, for it releases one's energies and makes one realise the powers at one's disposal. (AM, p. 187)

This says that violence plays an important role in almost all forms of anarchism and no exception is made for the author's own form. So I think that Gellner was entirely justified in reading the passage he quoted as an expression of Feyerabend's own view. But I also think that Feyerabend may be technically right in denying that he was asserting that violence is beneficial for the individual; for I think that he was doing his best to *insinuate* this idea but in a way that would allow him to wriggle out of responsibility for it if challenged. End of digression.

I have counted at least seven references by Feyerabend to my "letter to Santa Claus" (two of them in his 1976a). For instance, he writes (1978a, p. 166):

> If one asks a Popperian why on earth one should accept his standards he will answer that this is how science proceeds (. . . Lakatos), or that the standards are fruitful in the sense that they make us understand science (. . . Popper . . .), or that they lead to a science one "would like to have" (. . . Watkins in his "letter to Santa Claus").

He has "looked at every line of the position paper" (p. 153) but I wish that he had read them a little more carefully; for he makes a nonsense of what I was doing there. I did not, of course, propose as an aim for science what "we would like to have". What I did propose was that we should *begin* by erecting "a naively optimal ideal for science" and then ask which extant methodology retrieves the most of that naive ideal from the wreckage caused, chiefly, by Humean skepticism. I suggested that a naively optimal ideal would include both *certainty* and *ever deeper* (or even ultimate) *explanations*, and that these two aims pull in opposite directions. And I claimed that the Popperian methodology is able to capture more of that naive ideal by altogether discarding the goal of certainty, than does any methodology that hankers after something approaching certainty. (I now think that I understated my case: I should have claimed that a methodology that goes for certainty will capture *nothing* of that ideal, whereas one which goes for ever deeper explanations can capture *everything* except, of course, for the discarded certainty).

It still seems to me that the right way to try to find an aim for science that is both realistic and optimal, or neither over- nor underambitious, is to begin with a most utopian and nonviable ideal and then to consider in what ways it has to be cut back to make it viable and nonutopian. But, of course, this way of proceeding is automatically exposed to Feyerabend's unfailing one-upmanship method. After quoting my claim that the Popperian methodology preserves more of the Bacon-Descartes ideal than does any rival methodology, he comments: "The 'argument' assumes that the Bacon-Descartes ideal is worth preserving—which is the question at issue" (1978a, p. 176, note 16). I will not be so foolish as to argue for this assumption, for that ideal involves, among other things, truth and consistency, and we know what Feyerabend thinks of *these*.

Feyerabend's intellectual heroes change with bewildering rapidity. Not long ago Cohn-Bendit was at the top. Now, the "much maligned Bellarmine" (p. 172) is in the ascendant. For Feyerabend has developed a solicitude for theology ("the theory of God") and considers it a criticism of science that "science disturbs theology" (p. 172). Whether this support from this unexpected quarter will enable churchmen to sleep easier at night I do not know.

Postscript

Feyerabend was allowed a reply to this reply. He ended it on a saddened note: 'I have known the gentleman for now almost 30 years, I have participated with him in many amusing events, we fought some interesting battles together, I have often enjoyed talking to him, both on technical matters and on matters of "life" and have always profited from his observations. I do not understand why his present comments on my paper should be so far below his usual level of competence'. And now, unhappily, I get the last word. I will take it from the obituary I mentioned. I there said that Paul Feyerabend was one of the most gifted, colorful, original, and eccentric figures in postwar academic philosophy—irreverent, brilliant, outrageous, life-enhancing, unreliable and, for most who knew him, a loveable individual.

Notes

1. *The Independent*, March 4, 1994.
2. See Feyerabend (1995), chap. 7.
3. Feyerabend (1981), p. 47n.
4. I tried to chronicle this rift in my (1997) and to identify its propositional content in (forthcoming).
5. Feyerabend (1978c), p. 180.
6. Feyerabend 1978b, p. 390.

References

R. G. Colodny (ed.) (1965) *Beyond the Edge of Certainty*. Englewood Cliffs, N.J.: Prentice-Hall.
P. K. Feyerabend (1962) 'Explanation, Reduction and Empiricism', in H. Feigl and G. Maxwell (eds.), *Scientific Explanation, Space & Time, Minnesota Studies in the Philosophy of Science*, Vol. 3. Minneapolis: University of Minnesota Press. (Reprinted with additions in Feyerabend (1981), Vol. 1, Chap. 4).
P. K. Feyerabend (1965) 'Problems of Empiricism', in R. G. Colodny (ed.), 1965.
P. K. Feyerabend (1975) *Against Method*. London: New Left Books.

P. K. Feyerabend (1976a) 'On the Critique of Scientific Reason', in C. Howson (ed.), *Method and Appraisal in the Physical Sciences*. Cambridge: Cambridge University Press.

P. K. Feyerabend (1976b) 'Logic, Literacy and Professor Gellner', *The British Journal for the Philosophy of Science*, 27, pp. 381–391; reprinted in Feyerabend (1978c).

P. K. Feyerabend (1978a) 'In Defence of Aristotle: Comments on the Condition of Content Increase', in Radnitzky and Andersson (eds.), 1978.

P. K. Feyerabend (1978b) 'The Gong Show — Popperian Style', in Radnitzky and Andersson (eds.), 1978.

P. K. Feyerabend (1978c) *Science in a Free Society*. London: New Left Books.

P. K. Feyerabend (1981) *Philosophical Papers*, 2 Vols. Cambridge: Cambridge University Press.

P. K. Feyerabend (1995) *Killing Time*. Chicago: University of Chicago Press.

E. A. Gellner (1975) 'Beyond Truth and Falsehood', *The British Journal for the Philosophy of Science*, 26, pp. 331–342.

K. R. Popper (1940) 'What Is Dialectic?' *Mind*, 49, pp. 403–426; page references to Popper (1963), Chap. 15.

K. R. Popper (1957) 'The Aim of Science', *Ratio*, 1, pp. 24–35; reprinted in Popper (1972), Chap. 5.

K. R. Popper (1963) *Conjectures and Refutations*. London: Routledge and Kegan Paul.

K. R. Popper (1972) *Objective Knowledge: An Evolutionary Approach*. Oxford: Clarendon Press.

G. Radnitzky and G. Andersson (eds.) (1978) *Progress and Rationality in Science*. Dordrecht: Reidel.

J. W. N. Watkins (1966) Review of Colodny (ed.), (1965) *Philosophy*, 41, pp. 359–362.

J. W. N. Watkins (1971) 'CCR: A Refutation', *Philosophy*, 46, pp. 56–61.

J. W. N. Watkins (1978) 'Corroboration and the Problem of Content-Comparison', in Radnitzky and Andersson (eds.), 1978.

J. W. N. Watkins (1984) *Science and Scepticism*. Princeton: Princeton University Press, and London: Hutchinson.

J. W. N. Watkins (1997) 'Karl Raimund Popper, 1902–1994', in *Proceedings of the British Academy*, 1996 Lectures and Memoirs, 94, Oxford: Oxford University Press, pp. 645–684.

J. W. N. Watkins (forthcoming) 'The Propositional Content of the Popper-Lakatos Rift'.

6

Gonzalo Munévar

A *Réhabilitation* of Paul Feyerabend

Introduction

One of Paul Feyerabend's pet peeves was the way the history of philosophy had maligned Ernst Mach by treating him as the founding father of logical positivism. Feyerabend often spoke of writing a *réhabilitation* of Mach and did take a few beginning steps in that direction.[1] I largely agree with Feyerabend's appraisal of the historical misjudgment of Mach, but my purpose in this chapter is not to explore that issue: my intent is rather to write about Feyerabend as he would have about Mach. For it seems to me that few philosophers have been as misunderstood as Feyerabend himself.

I am particularly interested in discussing the following:

(1) Feyerabend's alleged claim that "anything goes";
(2) his alleged "incommensurability" thesis;
(3) his alleged relativism.

Before discussing Feyerabend's views, however, I would like to say that over his career Feyerabend changed his mind about a variety of subjects, sometimes drastically. He had a right to grow intellectually—he liked to remark—to try many approaches. That is the least we should have expected from such an able proponent of pluralism. This should be remembered as I give a sketch of Feyerabend's epistemology of science circa the publication of the first edition of *Against Method* (1975). As I go along I will try to point out those instances where Feyerabend changed his mind significantly.

One last preliminary point: what I am about to do could be seen as offering a *theory* about Feyerabend's philosophy, a systematic and consistent account of it. This he would have hated very much. He disliked theories of knowledge because he thought that knowledge, a complex part of human life, was always changing, as it should. He swore off relativism because relativism was a *theory* of knowledge.

It would thus not have pleased him to think that his "philosophical life" could be so easily "nailed down" in just one essay, that a "neat" account of it could so easily "freeze" what should have been fluid and open-ended.

Now, Feyerabend's position on the three issues I wish to discuss is grounded on his analysis of the history of science. It is from his understanding of the practice of science that much of his philosophy follows in a principled way, and it is thus to that understanding that I will first devote my attention.

1. Feyerabend's Analysis of the History of Science

It was Feyerabend's philosophical analysis of crucial episodes in the history of science that first led him to suggest his famous (or infamous) "Anything goes", his notion of incommensurability, and his epistemological relativity.[2] Of those crucial episodes, Feyerabend's favorite was Galileo's rebuttal of the formidable objections from physics, astronomy, philosophy, and religion against the motion of the Earth. Empiricist history of science had often painted Galileo as a patron saint of its main credo: that experience is the final judge of theory. There are disputes among the various forms of empiricism as to whether experience (the "facts") can justify a scientific theory, but in a crucial way most, if not all, agree that a theory in conflict with experience is not acceptable. Ironically, Galileo's own works gave Feyerabend his most powerful argument against the empiricist credo.

Among the many telling objections against the motion of the Earth, perhaps the Tower Argument presented the strongest of all. It goes as follows. Suppose that you let go of a stone from the top of a tall tower. If the Earth moves, by the time the stone hits the ground, the tower, being stuck in the Earth, will have moved considerably (the velocity of rotation of the Earth would have been calculated to be about a thousand miles per hour). Thus there should be a perceptible difference between the initial and final distances from stone to tower. But when we actually look, there is no difference at all! For the distance to remain constant, if the Earth did move, the stone would have to fall in a curved path, but we clearly see the stone fall straight down. Therefore the Earth does not move.

Faced with the Tower Argument what could Galileo say in defense of the Copernican view? He said that the stone does not fall straight down, no matter how clearly we see it. The stone only *seems* to move so; its real motion is far more complicated than that. But this makes no sense, people thought: motion is *observed* motion. Not so, said Galileo. Shared motion goes unobserved (motion is relative).

The reason why the stone keeps its distance from the tower is that its real motion has two components: the first is straight down, and we notice it, the second (which Feyerabend called circular inertia) is shared with the Earth, the tower, and the observer (us). That is why we do not notice it; but it is there all the same. Just as the tower moved laterally, so did the stone. When we are in an airplane flying smoothly we do not *perceive* that the passenger sitting next to us and the magazine on our lap are traveling at 650 miles an hour, even if we know they both are. We do perceive the motion of the stewardess up and down the aisle and of the drink spilling on the sleepy man to our left. We perceive those motions that we do not share, but fail to notice those that we do. Galileo made the same point by means of examples about ships, and in this manner he neutralized the objection against the Copernican view.

What conclusions did Feyerabend draw about the Tower Argument then? People noticed a phenomenon and *interpreted* it in what they thought was the most natural way, that is, the stone *moves* only straight down. It was this *natural interpretation* of the phenomenon, not the phenomenon itself, that contradicted the Copernican view. Galileo did away with the contradiction by providing a *different set of natural interpretations*. Galileo, then, constructed a new empirical basis! This new empirical basis, furthermore, is constituted by *a new theory of interpretation* congenial to Copernicus.[3]

What permitted the change of empirical basis was, in part, a change in theoretical assumptions. More specifically, Galileo changed the concept of motion. Motion had been supposed by his opponents, the Aristotelians, to be only observed motion. In the jargon of this century, we might say that the Aristotelians had an operationalist concept of motion (i.e., that a phenomenon would count as motion only if it could be expressed in terms of observable changes). But Galileo introduced into motion components that could not be observed. And in the particular case of the Tower Argument, one of those components was circular inertia, which in addition to being highly theoretical (and ad hoc) was quickly abandoned after the Copernicans won the scientific revolution.

Philosophers have often praised Galileo for preferring his eyes to his Aristotle,[4] but Feyerabend's analysis shows how misguided that praise has been. Galileo was a Copernican. The central thesis of Copernicus stated that the Earth was one more planet in orbit around the Sun. The evidence of the eyes, unfortunately, refuted this Copernican thesis. As we well know now, the Earth and Venus are sometimes on the same side of the Sun, and thus rather close to each other; and sometimes they are on opposite sides of the Sun and thus very far from each other. The same can be said for Mars, except that Mars gets much farther from the Earth. It seems commonsensical then that Venus and Mars should look brighter when they are close and dimmer when they are far. But the magnitude of Venus barely changes, and that of Mars does not change as much as it should.

Nevertheless Galileo's admiration for Copernicus did not decrease, even though Copernicus, in Galileo's words, "with reason as his guide . . . resolutely continued to affirm what sensible experience seemed to contradict".[5] Reason, it appears, can overturn the verdict of experience: "there is no limit to my astonishment", Galileo writes, "when I reflect that Aristarchus and Copernicus were able to make reason so conquer sense that, in defiance of the latter, the former became mistress of their belief" (ibid.). In his own case, Galileo says, he came upon "the existence of a superior and better sense than natural and common sense", the telescope, which then joins "forces with reason" (ibid., p. 103). Fortunately enough, his telescopic observations of Mars gave magnitudes much more in agreement with Copernicus's thesis. And the telescope showed that Venus has phases. When Venus is farthest from us, we see its full face illuminated by the Sun. When it is closest, most of the face that it shows to us is dark. Thus the amount of light that reaches us from Venus remains constant enough for our eye to perceive little change in magnitude.

In this situation we find an outright conflict between natural and artificial sense. Galileo surely saw it that way. He resolved the conflict by *denying the testimony of his eyes* and siding with the sense that agreed with Copernicus, the telescope. This conflict between senses should dispel any illusions to the effect that scientific instruments merely amplify and sharpen our senses. But some empiricists may feel relieved that it was telescopic *experience* after all that decided the matter. Unfortunately, such relief is unjustified unless they can show that the telescopic experience was without question superior to that of the eye, as far as the observers of that day were concerned, *and* that the choice of the telescope over the eye does not require theoretical assumptions.

Empiricists will have difficulties on both counts. Kepler, for example, wrote to Galileo:

> I do not want to hide it from you that quite a few Italians have sent letters to Prague asserting that they could not see those stars (the moons of Jupiter) with your own telescope. I ask myself how it can be that so many deny the phenomenon, including those who use a telescope. Now, if I consider what occasionally happens to me, then I do not at all regard it as impossible that a single person may see what thousands are unable to see. (ibid., p. 124)

Feyerabend had many interesting things to say about the Italian observers' troubles with Galileo's telescope, but for the sake of brevity let me discuss only the second difficulty.

The second, and major, difficulty for empiricism is that Galileo's trust in the telescope required the granting of several *theoretical* assumptions. Images from the heavens would travel immense distances, enter a different medium upon hitting the Earth's atmosphere,[6] work their way through the telescope, and finally be handled by a brain that had never perceived anything like them. To be assured that those images were not significantly distorted, Galileo needed supporting theories about optics, about the nature of light, about the atmosphere, about the interaction between light and a variety of gases, about the telescope, and about perception. We may realize, then, that it was not Galileo's telescopic *observations* that challenged the geocentric view of the universe, but his observations together with a host of assumptions from many supporting or auxiliary sciences. The crucial question was: could experience alone have reconciled the magnitudes of the planets with Copernicus's thesis? If by experience we mean sensory experience, the answer is no. If we allow telescopic experience, then we should remember that such experience could be taken as reliable only if interpreted on the basis of certain theories. The answer again is no.

To make matters worse, most of the auxiliary sciences in question were not within Galileo's reach. Some of them took hundreds of years of development before they could fully back Galileo's hunches. Thus to a good empiricist of the day, many of Galileo's theoretical assumptions should have seemed unwarranted.

Galileo's scientific instincts may seem unerring to us, with the benefit of hindsight, but that is a different issue. Nevertheless, let us turn the tables and see why we should recommend to Galileo to act as he did. The same analysis that leads to the realization that Galileo made theoretical assumptions applies to the eye as much as it does to the telescope. Visual perception is a complex process in which the brain takes into account not just the "input" from the retina but also from the inner ear and hundreds of skeletal muscles (to determine the position of the body) and from the other senses, as well as from memory and imagination. Think of how vague images suddenly come into focus when we smell the particular scent of a flower in a forest or hear the growling of a dog in a dark street.

The working of perception in these and other complex ways is the result of the history of adaptations of the brains of our ancestors to a variety of environments. The extent to which the senses can be "trusted" is thus not a matter for philosophy alone to determine. Psychology tells us of the richness and complexity of perception, neuroscience may help reveal the structures that make such richness and complexity possible, and evolutionary biology may explain how those structures arose and may also give us clues about where they apply.

Trying to decide between the eye and the telescope is a very complicated affair that involves a host of theoretical assumptions. Of course, Galileo did not know anything about neuroscience, let alone evolutionary biology. But he did realize that his opponents' assumption about the reliability of the senses had to be backed up by some view of the relation between the world and the senses. Indeed that view was Aristotle's theory of perception, according to which an undisturbed mind will take on the "form" of an object as long as that form has traveled through an undisturbed medium. That is how a normal observer under normal conditions would gain knowledge of the world. Thus behind the clash of the senses there was a clash of theories, whether those theories were explicit or not. Galileo chose the theoretical direction that promised him the most exciting discoveries.

It is important to notice that Galileo did not merely take methodological shortcuts. It is not as if his hunches had led him more quickly to results that more patient methodologists would have achieved eventually. Not at all. If method requires the priority of experience, method would have forever closed the door on a view that could not be established without overthrowing accepted experience. If, in pursuing a theory that had been refuted by experience, Galileo committed a sin against science and philosophy, we should love not only the sinner but the sin.

This is not to say, however, that theory is always overturning the verdict of experience. Nevertheless, it is to say that it *may*. Thus we should not be startled upon hearing that some brilliant scientist showed little concern about the results of apparently crucial experiments or observations. When Einstein was asked what he would have thought if an important experiment had disconfirmed his theory of general relativity, he answered, "Then I would have to be sorry for dear God. The theory is correct".[7]

I trust this sketch suffices to show how Feyerabend's analysis of the history of science led him to uncover some very important limitations of the empiricist credo. Of course, empiricists did not take kindly to the use that Feyerabend, and Kuhn, made of history. Notice, however, that they could not dismiss history (i.e., practice) on the hitherto successful grounds that the epistemology of science deals not with how science is done but with how it ought to be done. For what history showed us, through Kuhn and Feyerabend, was that in order to achieve scientific success, as proclaimed by empiricists, scientists sometimes had to violate the methodological prescriptions of those very empiricists.

An important empiricist counterattack concerns the nature of the evidence that Kuhn and Feyerabend have used against the standard views in philosophy of science. Kuhn and Feyerabend have used history to make their points. But how do they know that history can be trusted to that extent? How is it that the "facts" of physics, astronomy, and chemistry can be overthrown but we must show reverence for the "facts" of history? Kuhn and Feyerabend cannot have it both ways. Indeed, it seems reasonable to suppose that physics is more reliable than history. Thus if Kuhn and Feyerabend's thesis against the theory/observation distinction is granted, their evidence is not worth all that much.

Several important points must be considered in reply. The first is that when the objection is aimed against Feyerabend it fails to take into account the structure of his argument. To see this more clearly it pays to consider first a similar objection. According to this other objection, Feyerabend argues against reason in science. But in order to establish his conclusion, Feyerabend has to use reason. If reason is no good, Feyerabend's means for establishing his conclusion are no good either. Thus Feyerabend must be committed to the correctness of reason. Unfortunately for him, this consequence invalidates his conclusion.

What this objection misses completely is that Feyerabend does not need to be committed to the correctness of reason, whether as rational argument or methodological rules. His argument is a reductio ad absurdum, and a rather simple one at that. In a *reductio* one assumes for the sake of argument the opponent's position and then derives a conclusion unacceptable to that opponent (that is, reduces his position to absurdity). Feyerabend drives this point home time and again. He says, for example,

> Always remember that the demonstrations and the rhetorics used do not express any "deep convictions" of mine. They merely show how easy it is to lead people by the nose in a rational way. An Anarchist is like an undercover agent who plays the game of Reason in order to undercut the authority of Reason. (AM, pp. 32–33)

Of course, if reason is no good Feyerabend has no argument. But he does not need one, since the conclusion is already established.

Now we can consider the initial objection. How can Feyerabend trust the "facts" of history after undermining the apparently far more solid "facts" of physics, astronomy, chemistry, and the like? Furthermore, is Feyerabend now arguing that methodologies are refuted by facts and thus have to go? But did he not argue against this very way of dispensing with views? (cf. his arguments against falsificationism). Once again Feyerabend takes the cue from his opponents:

> *They* prefer Galileo to Aristotle. *They* say that the transition Aristotle/Galileo is a step in the right direction. I only add that this step not only was *not achieved*, but *could not have been achieved* with the methods favoured by them. But does not this argument involve highly complex statements concerning facts, tendencies,

physical and historical possibilities? Of course it does, but note that I am not committed to asserting their truth.[8]

Once again, Feyerabend's argument is a reductio. His aim is not to establish the truth of propositions but to make his opponent change his mind (or at least give him pause to consider why he should not change it). To achieve this end, he says,

> I provide him with statements such as 'no single theory ever agrees with all the known facts in its domain'. I use such statements because I assume that being a rationalist he will be affected by them in a predictable way. He will compare them with what he regards as relevant evidence, for example, he will look up records of experiments. This activity, combined with his rationalistic ideology will in the end cause him 'to accept them as true' (this is how he will describe the matter) and so he will perceive a difficulty for some of his favourite methodologies. (SFS, p. 143)

The assumptions that Feyerabend makes about the efficacy of his rhetorical devices, as well as his motivation, should be of no concern to the rationalist.

> All he *needs* to consider, all he is *permitted* to consider is how the statements surrounding the case studies in my book are related to each other and to the historical material and whether they can be read as an argument in *his* sense. I admit that my procedure succeeds by manipulating the rationalist but note that I manipulate him in a way in which he wants to be manipulated and constantly manipulates *others*: I provide him with material which interpreted in accordance with the rationalistic code creates difficulties for views he holds. Do *I* have to interpret the material as he does? Do *I* have to 'take it seriously'? Certainly not, for the motivation behind an argument does not affect its rationality and is therefore not subjected to any restriction. (ibid.)

It is not clear, by the way, that Kuhn also intends to offer a *reductio*. But the objection does not work against him either. In the first place, Kuhn does not claim that conflict with historical facts makes it obligatory to abandon traditional epistemology of science:

> [The contrary historical facts] by themselves . . . cannot and will not falsify that philosophical theory, for its defenders will do what we have already seen scientists doing when confronted by anomaly. They will devise numerous articulations and *ad hoc* modifications of their theory in order to eliminate any apparent conflict.[9]

Such historical anomalies can then at best "help to create a crisis or, more accurately, to reinforce one that is already very much in existence" (ibid.). Their main contribution to the epistemology of science is that they make possible "the emergence of a new and different analysis of science within which they are no longer a source of trouble" (ibid.).

Apart from the analysis of the reasoning involved, a key point at issue is the overthrow of the empirical basis. If it can be done for physics it can surely be done for history as well. But then, views of science that take such historical facts seriously can have the rug pulled from under their feet no less than theories in physics or astronomy. Nevertheless we do not have much of an objection. After all, the task at hand would require the overthrow of the present sociohistorical empirical basis that has been used by Kuhn and Feyerabend. Theories of science are thus in the same boat with scientific theories: they can be replaced by alternatives that change the "facts" on them. But that possibility does not diminish their worth. There will be no decisive objection along the present lines, therefore, unless the empirical basis of this view of science is actually overthrown.

If Feyerabend is right, all methodological principles have exceptions (progress requires their occasional violation). This result alone has made many rationalists conclude that science would then be ruled by anarchy. Feyerabend seems to accept this interpretation when he proclaims that "science is an essentially anarchistic enterprise" (AM, p. 17), but his emphasis is not that science falls short of what it ought to be, but rather that what it "ought to be" is nothing but an unwarranted imposition by philosophers on an enterprise that investigates a complex medium. As he puts it, "A complex medium containing surprising and unforeseen developments demands complex procedures and defies analysis on the basis of rules which have been set up in advance and without regard to the ever-changing conditions of history" (AM, p. 18). A methodologically correct science would be an inferior science.

2. "Anything Goes"

When Feyerabend claims that "the only principle that does not inhibit progress is: anything goes" (AM, p. 23), he is gunning for the philosophy of science. As he says, "*anarchism*, while perhaps not the most attractive *political* philosophy, is certainly excellent medicine for *epistemology*, and for the *philosophy of science*" (AM, p. 17). He arrives at this result on the basis of his examination of historical episodes and of his analysis of the relation between idea and action. This is by no means a negative, skeptical philosophy, as we will see in this section. "Theoretical anarchism", he says, "is more humanitarian and more likely to encourage progress than its law-and-order alternatives" (ibid.).

Feyerabend's aim is to promote the advantages of pluralism in science, a course of thought that conflicts with Kuhn's ideas, for Kuhn argued that the tenacious commitment to one point of view is the best way to ensure progress in science. An examination of this difference with Kuhn will shed light on some important aspects of Feyerabend's philosophy.

Now, Kuhn correctly points out that a comprehensive view is abandoned not because it has anomalies, but because it is replaced by an alternative. Anomalies thus do not refute a paradigm, but they may bring about a crisis if they are thought to be important enough (for then the failure to assimilate them gains great significance). No anomaly, however, is as important as one which a competitor claims to have explained—no anomaly, that is, accentuates more the loss of confidence in the paradigm. The reason is that, as Kuhn explains, a paradigm is accepted on the promise of future performance, on the promise, that is, that it will prove the best way to conceive of the world.[10] When a competing would-be paradigm seems to be doing better, our faith in the *promise* of our anomaly-besieged paradigm may well falter. Thus, Feyerabend thought, we will create more crises, and therefore more fruitful change (in Kuhn's own terms) by providing a mechanism to strengthen the anomalies. To accomplish this goal, science should be organized so as to require the *continuous generation of alternatives*. This Feyerabend calls the *principle of proliferation*.[11]

Expectation can cut two ways: it permits us to see, but it may also keep us from seeing. Can we do better? Yes, if we have more than one set of expectations to draw from. That is Feyerabend's point. Of course, it may be difficult for the same individual to hold alternative sets of expectations. On the other hand, in a discipline where proliferation is not ruled out, some individuals may develop a different worldview, and may thus be in a position to point out areas of difficulty that the other members of the discipline may tend to overlook. That is the first step. The second step is that where one side may face contrary facts or fuzzy pictures, the rival may claim consistency with observation (within its system) or clear pictures. At that stage it may be far more difficult for the practitioners of the standard view to regard the anomalies as mere oddities that may be taken care of at a later date. The failure to articulate the paradigm may then acquire great significance precisely because of the pressing possibility that the paradigm's categories may not provide the best box into which to file the experience in question. And this suspicion of doom grows larger in proportion to the ease demonstrated by the rival view.

Two resolutions are then made available: (1) the defenders of the standard view do manage to assimilate the anomaly, or (2) a crisis ensues that may result in a paradigm shift. Either resolution is prompted by the presence of a rival. Without the competition by that rival there would have been no motivation for worry in the first place. That is, even according to Kuhn's own specifications, the scientific discipline would not have done as well. The point is not really all that intricate. It is possible to discover one's own faults through introspection. But the great majority of us learn of our faults thanks to all those around us who are plainly delighted to point them out.

On the other hand, from Kuhn we have learned—against falsificationism— that a view should not be dropped merely because some of its predictions fail. A view should be given time to develop, to bear out its original promise. Thus commitment to a view is crucial to science (Feyerabend calls this the *principle of tenacity*). Where Kuhn goes too far is in his insistence that *the entire discipline* be dedicated to only one view. It is more fruitful for science, Feyerabend says, to have several competing groups working on those ideas that they find particularly promising. It is also a more humanitarian outlook, for instead of authoritarian enforcement of a dominant view, it allows for the individual scientist's pursuit of happiness through his scientific work.

Feyerabend makes use of two simple notions. First, people work best at those things they like best. Thus scientists should be encouraged to develop those views of nature that for any reason have caught their fancy. Second, the quality of work improves when strong challenge points the way in the direction that requires improvement. Without such challenge we become complacent, and even if we want to keep on our toes we may fail to see the flaws in our pet theories because we are too close to them, or because we lack enough imagination, or for many other reasons. Thus competition aids us in our pursuit of excellence.

The principle of proliferation, and within its action, the principle of tenacity, leads to greater human happiness. Those two principles also create the conditions for fruitful change and improvement. Thus both humanity and science are the better for their presence. Kuhn's insistence on the restriction of the principle of tenacity to only one view is the sort of dogmatism that gets in the way of what Kuhn himself considers advantageous. On the other hand, the falsificationist's rigid quest for rejection robs science of the depth tenacity offers to it—without contributing to individual happiness, for scientists are obliged to give up what may well intrigue them. It seems that Feyerabend is offering us the opportunity to have our scientific cake and eat it too.

Feyerabend does not claim that science in fact progresses. He leaves the evaluation of progress up to each individual (or each individual group). He only tells the different parties how they can best bring about their own goals by adhering to the principle of tenacity and permitting the functioning of the principle of proliferation. If by continuous confrontation with competing views a scientist is spurred to improve his view—where improvement is measured by his own standards—then his work has profited. If, on the other hand, he finds that the competition has gotten the best of him—again, for whatever reasons—he abandons his view and takes up the other. In this case he has also profited. The appropriate mix of the principles of tenacity and proliferation increases the scientist's chance of scientific profit *as seen from his own methodological point of view.*

This is just one variation on Feyerabend's grand theme of scientific anarchy. His tune against Kuhn's dogmatism is an adaptation of a general approach that he developed against the standard methodologies. As we saw earlier, Feyerabend shows that even the most obvious methodologies have limitations. For empiricism it makes no sense to use hypotheses that, say, contradict well-confirmed theories, and even less sense to use hypotheses that contradict well-established experimental results. But Feyerabend argues that such hypotheses may be used to the great advantage of science. Can we really advance science by proceeding counterinductively?

Yes. The reason, as we may have surmised from Feyerabend's case against Kuhn's dogmatism, is that those counterinductive hypotheses give us evidence that cannot be obtained in any other way. Prejudice is often discovered not by analysis but by contrast. If, as we have seen, every fact is already viewed in a certain way, and to progress often requires viewing facts in a different way, then we simply need alternative ways of seeing. As for the conflict between those counterinductive hypotheses and the facts—and it is that conflict that presumably makes them counterinductive—we should remember that no theory ever agrees with all the facts in its domain (cf. Kuhn's account of how a paradigm is a promise of results and not a collection of them). If such conflict is grounds for throwing out a theory, then we should throw out all theories. The main reason for not trembling in the shadow of the facts is that facts are constituted, in part, by older ideologies, and thus a clash between facts and theories may actually be an indication of progress—an indication that our probe is coming into contact with some of the principles assumed in familiar observational notions.

It is often said that we cannot step outside science to see whether it represents the world. This simple point is supposed to dog the idea that truth is correspondence to reality. And maybe it does. But we may still observe the relationship between our science and the world by comparing our science with an alternative interpretation of what the world is like. As Feyerabend says, "We *need a dream-world in order to discover the features of the real world we think we inhabit* (and which may actually be just another dream-world)" (*AM*, p. 32). In this Feyerabend echoes John Stuart Mill. If our present views are right, by criticizing them from another vantage point, we come to understand them better. And if they are not right, we gain the opportunity to replace them.

If this is so, however, we come to realize that any idea, no matter how ancient or absurd, is capable of improving our knowledge. This sounds at first preposterous. For example, we finally got rid of all that Aristotelian nonsense in science. Why bring it back? But then, many of the central ideas of modern science were once considered preposterous. Consider, to name only three: heliocentrism, held by Aristarchus; atomism, held by Democritus; and evolution, held by Lamarck, and before him by even more disreputable characters. Of course, the modern versions of those ideas are quite different. But the fact of the matter is that thinkers like Copernicus, Dalton, and Darwin found promise in those discredited ideas and took the trouble to develop them. To those thinkers we owe in large part the glory of modern science. Here we find in action both the principles of proliferation and of tenacity.

The operation of these last two principles makes science appear far more "sloppy" and "irrational" than its methodological image. But as we have seen, the attempt to make science conform to that methodological image, the attempt, that is, "to make science more 'rational' and more precise is bound to wipe it out" (*AM*, p. 179). For, as Feyerabend argues, "What appears as 'sloppiness', 'chaos' or opportunism when compared with such [an image] has a most important function in the development of those very theories which we today regard as essential parts of our knowledge of nature" (ibid.). But then from the methodologist's point of view nothing can be ruled out, *anything goes*. And if methodology is equated with reason, science is and must be an irrational enterprise. As Einstein once put it:

> The external conditions which are set for [the scientist] by the facts of experience do not permit him to let himself be too much restricted, in the construction of his conceptual world, by the adherence to an epistemological system. He therefore must appear to the systematic epistemologist as a type of unscrupulous opportunist.[12]

The situation is then as follows. According to the rationalist, alias methodologist, alias systematic epistemologist, certain events in the history of science constitute progress. But for those events to come about some scientists have to be opportunistic enough to adopt whatever procedure seems to fit the occasion. This means that even the best methodological rules *must* be violated from time to time. But this inherent limitation of all rules implies that nothing can be ruled out once and for all. To a methodologist this amounts to an admission that *anything goes*. Therefore, *from the methodologist's point of view*, anarchy will occasionally be essential to science.

This principle of anarchy is the only principle that does not inhibit progress, according to Feyerabend. But can it really be the case that anything goes in science? Although Feyerabend sometimes points out how ideas that today are considered preposterous have much to offer, he does this partly for rhetorical purposes and partly to annoy his educated opponents. For he does not argue that all ideas and procedures fit all circumstances equally well. On the contrary, in the case of Galileo and others he illustrates how some specific ideas and procedures were particularly helpful. *Anything goes* not from his point of view, but from the point of view of one who thinks that only certain ideas or procedures are admissible. Anarchy is thus in the eye of the rationalist beholder, a point that Feyerabend expresses with emphasis: " . . . '*anything goes' does not express any conviction of mine, it is a jocular summary of the predicament of the rationalist*" (SFS, p. 188).

If anarchy means ignoring the rationalist's rules from time to time, then science requires anarchy. This is not to say that the rules never apply: if anything goes, reason sometimes goes too. Nor is this simply a bromide to the effect that since science is a human activity it *cannot* be perfectly rational. The point is rather that science *must not* be perfectly rational (in the rationalist sense that equates rationality with adherence to rules), if it is to achieve progress.

Nor is Feyerabend putting forth a new methodology (e.g., a counterinductive methodology such as "advance theories inconsistent with the facts"). What we cannot do is precisely to specify in advance whether the inductive or the counterinductive rules will apply. This would not change even if somehow we could foresee the context that the scientist will face. For the different choices that he may make when faced with a new context may themselves change that context in many different ways. When Dalton imported into chemistry his notions of simple ratios, he changed as a consequence the concepts of mixture and compound, as well as the standards of chemical explanation.

It is nevertheless uncanny that Feyerabend seems to be blessed with such special knowledge about what is good and what is bad in science. He certainly seems to have no qualms about using words like *progress, advance, improvement*, and so on. Is there perhaps at least a rational way of appraising episodes in the history of science? But Feyerabend does not have to figure out whether some episode or other really constituted progress. His concern is of an altogether different sort. As he says

> *Everyone can read the terms in his own way* and in accordance with the tradition to which he belongs. Thus for an empiricist, 'progress' will mean transition to a theory that provides direct empirical tests for most of its basic assumptions. Some people believe the quantum theory to be a theory of this kind. For others, "progress" may mean unification and harmony, perhaps even at the expense of empirical adequacy. This is how Einstein viewed the general theory of relativity. *And*

> *my thesis is that anarchism helps to achieve progress in any one of the senses one cares to choose. Even a law-and-order science will succeed only if anarchistic moves are occasionally allowed to take place. (AM, p. 27)*

Has Feyerabend, then, done away with rationality in science? My own inclination, which I have followed elsewhere,[13] is to take to heart Kuhn's advice:

> If history or any other empirical discipline leads us to believe that the development of science depends essentially on behavior that we have thought to be irrational, then we should conclude not that science is irrational, but that our notion of rationality needs adjustment here and there.[14]

Those adjustments, given Feyerabend's results, are bound to be quite dramatic. It is difficult to see how philosophy of science could return to the days before *Against Method*.

3. Incommensurability

Feyerabend and Kuhn's claim that in times of radical scientific change experience may change, as we saw in the case of Galileo, had many other profound consequences in the philosophy of science. If this claim is correct, the growth of science is not cumulative, for there is a sense in which science may begin anew in times of revolution. This result was met with dismay by many philosophers of science. According to these philosophers, rationality in science requires that changes of points of view may be justified. And this justification must be made by showing that the winner is superior to the loser. The only "scientific" way to show this superiority was by comparing the views in question with experience: the "facts" were the common measuring standards by which the worth of theories was to be determined. But if we may have different sets of facts before and after the change, we can no longer rely on common standards, and thus we cannot show that the new view is superior to the old. Therefore, if Kuhn and Feyerabend are correct it cannot be shown that science is a rational enterprise. This is the problem created by the thesis of *incommensurability*, which, strictly speaking, means "lack of a common measure".

It is easy to see why this problem comes very straightforwardly out of Feyerabend's analysis of the history of science, and we can also see why it should become a headache for an empiricist epistemology. Analytic philosophers, however, tended to see this problem in linguistic terms. Such a linguistic turn was not unexpected. Given that it was common to think of theories as sets of (theoretical) sentences, or theoretical "languages" of some sort, science was expected to grow by linguistic accumulation. If Newton's theory was replaced by Einstein's, then it must be subsumed and explained under Einstein's. And to say that Einstein's theory explains Newton's is to say that Newton's must be derivable from Einstein's. Explanation was logical reduction, and of course logical reduction requires logical derivation. But unfortunately logical derivation fails if the common terms change meaning from the premises (the explanans) to the conclusion (the explanandum), which was exactly what Feyerabend also claimed.[15]

Analytic philosophers also feared that the failure of logical derivation would amount to a failure of communication of the following sort: according to Feyerabend (and Kuhn as well), the meaning of scientific terms is holistic, that is, it depends on the entire theory in which it plays a role. When the theory changes, the meaning of the terms employed changes also. Across theory change, then, scientists are speaking different languages. From this it supposedly follows that scientists who hold competing points of view (old versus new Kuhnian paradigms, for example) will fail to agree because they are bound to misunderstand each other.[16]

To be fair, Kuhn did speak of how practitioners of the old paradigm might not be able to see the "evidence" presented in support of an alternative as evidence at all (SSR, p. 94) and Feyerabend did speak of holistic meaning. Also, both philosophers explain how communication was possible across theoretical boundaries (you could learn both "languages", for example).

Nonetheless, my purpose in this section is to argue that the problem of incommensurability exists quite apart from any one particular theory of meaning. Feyerabend's holistic theory, the target of so many lectures and publications, is simply irrelevant, as he himself pointed out. In fact, he belittled the entire linguistic discussion:

> Putnam creates the impression that I am mainly interested in meanings and that I am eager to find change where others see stability. This is not so. As far as I am concerned, even the most detailed conversation about meanings belongs in the gossip columns and have no place in the theory of knowledge. This is true even in those cases where meanings are invoked to force a decision about some different matter. For even here their only function is to conceal some dogmatic statement which would not be accepted, if presented by itself, and without the chatter of semantic discussion.[17]

I will begin with a discussion of whether the transition from Newton to Einstein was cumulative, continue with an account of how the question of meaning crept in, and finish with an argument for the irrelevance of (nontrivial) semantic theories.

On grounds of economy, let me deal with the first point by reminding the reader of Kuhn's remarks concerning the Newton-Einstein transition. According to Kuhn, "Einstein's theory can be accepted only with the recognition that Newton's was wrong" (SSR, p. 98). But Newton's physics is widely thought to have a limited validity and is in fact still much in use by engineers and even many scientists. What Einstein did, according to Kuhn's critics, was to draw the proper boundaries for classical mechanics. Within those boundaries Newton's theory is essentially correct. Furthermore, taken within those boundaries (e.g., low velocities) it can be shown to be derived from relativistic mechanics. Therefore Kuhn is wrong: this century's revolution in physics did not replace what preceded it but rather added to it. Knowledge in physics has grown by accumulation.

This objection seems odd in light of the many baffling aspects of Einstein's special theory of relativity (to say nothing of his general theory). It is true, of course, that the two theories yield similar (although not the same) numerical values for mass, length, etc., at low velocities. Nevertheless, when the critics say, as they are obliged to, that Newtonians were unscientific if they thought that Newtonian theory applied at velocities close to that of light, those critics forget that "the price of significant scientific advance is a commitment that runs the risk of being wrong" (SSR, p. 101). Such a restriction, Kuhn points out, "forbids the scientist to rely upon a theory in his own research whenever that research enters an area or seeks a degree of precision for which past practice with the theory offers no precedent" (SSR, p. 100). The result of accepting it "would be the end of the research through which science may develop further" (ibid.). As Kuhn argues, to save Newtonian physics in this manner (from being overthrown) would permit us to save practically any theory that has been held by serious scientists, including the notorious phlogiston theory, for within the limits of its validity it is still valid (of course).

Moreover, the alleged derivation is spurious. A necessary condition for the validity of any derivation is that the terms preserve their meanings throughout (otherwise we commit the logical fallacy of equivocation). Both classical and relativistic mechanics employ concepts of space, time, mass, etc. But according to Kuhn, "The physical referents of [the] Einsteinian are by no means identical with those of the Newtonian concepts that bear the same name" (SSR, p. 102). For example, Newtonian mass is conserved, Einsteinian is convertible with energy.

In a similar fashion, Feyerabend also questions the attempt to identify classical mass with relativistic *rest* mass. And relativistic length, Feyerabend adds, "involves an element that is absent from the classical concept and is in principle excluded from it. It involves the *relative velocity* of the object concerned in some reference system" ('Consolations for the Specialist', p. 221). He concludes that "different magnitudes based on different concepts may give identical values on their respective scales without ceasing to be different magnitudes" (ibid., pp. 221–222). Where the terms involved have different meanings, then, no derivation is possible. Since this situation will occur in all those cases where a new comprehensive view of the world is born out of conflict, scientific revolutions are marked by the incommensurability of the theories (or paradigms) involved.

Some critics have suggested that it is possible to provide a "common dictionary", so it is possible to compare, say, the predictive successes of two competing views of the world. And there is a clear sense in which that is indeed possible, but this acknowledgment does not affect Feyerabend's view. In fact, the present example provides an illustration. The laws that give such good agreement in numbers with Einstein's at low velocities are not really Newton's laws, as he would have understood them, but rather a relativistic version of them. They are Newton's laws transposed into Einstein's special theory of relativity with limits and parameters on them that would have been inconceivable to Newton himself. If we were to take those laws, and Einstein's as well, as mere instruments for the management of data, the meanings of the crucial terms might remain the same. But those terms acquire meaning for the scientist only as his training makes it second nature for him to attach them to the world in ways guided by the view of the world accepted by his discipline. As long as the way a term is used has nothing to do with its meaning, then we do not have this semantic problem of incommensurability. On the other hand, if we do think that the use of the term has something to do with its meaning, then we must consider how our view of the world influences that use.

Even this much commitment to "holism" can be eschewed by Feyerabend. There were at the time two recurrent themes in analytic philosophy of science with respect to the meaning of scientific terms. At one extreme was Bridgman's operationism, according to which the meanings of scientific terms should be determined by operations (e.g., length is defined by specified ways of measuring: by the use of rods, by triangulations, or by bouncing radar signals off an object).[18] At the other extreme was the view, favored by Hempel, that scientific terms have what he called "systematic import", which means roughly that the meaning of a term depends in part on the role it plays in a theory—meaning is therefore affected by its relations to other terms also used in the theory.[19] Whatever extreme we tend to favor, or a suitable combination of both, we cannot avoid incommensurability. When we change paradigms (to stick to Kuhn's terminology) we obviously may change the relevant "operations", for the new paradigm may offer new and different experimental and instrumental commitments. And if we are inclined toward Hempel's view, it seems that drastic theoretical change may surely lead to change in meaning as well.

There was also a fixation in some circles that the meaning of "theoretical terms" had to be determined by prior "observational terms" (this is inductivism imported into semantics, i.e., the equivalent of "theories have to be derived from the facts"). If this view were correct, the meaning of observational terms would be prior and thus could not be altered by changes in the meaning of theoretical terms. It seems, however, that among the many assumptions made in such an argument, most of them very questionable, special prominence must be given to the distinction between theory and observation. But if this view of meaning needs such a distinction to even get off the ground, to advance it against those who question the distinction is to beg the question. Furthermore, we have seen how Feyerabend's analysis of the history of science demolishes such a distinction.

It is very important that we understand how far the analytic philosophers' accounts are from Feyerabend's notion of incommensurability. "Mere difference of meanings", he says, "does not yet lead to incommensurability in my sense".[20] Feyerabend's point was that Einstein's use of the terms *mass* and *length* precludes Newton's. According to Feyerabend, to say that Einstein's and Newton's theories (or paradigms) are incommensurable is to say that the principles for constructing concepts in one *suspend* the principles of the other. That is, the relativistic ideas used to understand mass, length, and time do not permit the use of the alternative Newtonian ideas. And surely, since concepts like "mass", "length", and "time" are used to determine what counts as facts in physics, we may conclude that presenting Einstein's facts means suspending principles assumed in the construction of Newtonian facts (AM, pp. 267–285).

This last way of putting the matter can bring us out of an issue over which philosophers have burned up much slobber for little profit. Notice how it returns us to the starting point of our discussion of this issue: incommensurability simply means that there are no common standards of measurement, that is, that there may be no common sets of facts to judge one theory or paradigm superior to another. If universal principles in a theory suspend the principles (of concept-formation) in a second, we realize that the empirical basis of the first is different from that of the second. But when we put the matter this way, we realize that we are just talking about the possibility of overthrowing the empirical basis. And this we have already observed in section 1 without semantic mirrors.

Although critics often concentrate their attacks on what they call Kuhn and Feyerabend's "holistic theory of meaning", neither Kuhn nor Feyerabend need commit themselves to any particular theory of meaning beyond the trivial point that the meaning of scientific terms is *somehow* connected with how scientists use them. The problem of incommensurability is an epistemological consequence of their analysis of the history of science, not of semantical idiosyncrasies on their part. Of course, the meaning of scientific terms does change from time to time, but that is no longer of great philosophical importance. It was so once, when analytic philosophy of science favored the view that logical derivation was an essential component of scientific explanation and that logical reduction was the right model for the growth of scientific knowledge. Feyerabend put the matter in perspective thus:

> I should add that incommensurability is a difficulty for philosophers, not for scientists. Philosophers insist on stability of meaning throughout an argument while scientists, being aware that 'speaking a language or explaining a situation means both following rules and changing them' . . . are experts in the art of arguing across lines which philosophers regard as insuperable boundaries of discourse. (*Farewell to Reason*, p. 272)

4. Relativism

I do not mean to deny that an air of paradox surrounds the issue of incommensurability and that a rather distinct odor of skepticism clings to it. If our paradigm or basic theory determines, among other things, the constituents of the world, one may say that "the world changes when basic theory changes". Or as Kuhn puts it: "after a revolution scientists are responding to a different world" (*SSR*, p. 111). "It is rather", he adds, "as if the professional community had been suddenly transported to another planet where familiar objects are seen in a different light and are joined by unfamiliar ones as well" (ibid.).

There are some who find this an exciting discovery about the nature of science. But others feel that we are confronted with a new skeptical problem: if the furniture of the world is relative to our worldview, then a revolution in science may well bring about a change of furniture. This is a prospect that most philosophers would find repugnant. Many of them did argue that even if scientific terms change meanings the important thing is that they still *refer to* the same objects. But on this point Feyerabend can be most irksome:

> [W]e certainly cannot assume that two incommensurable theories deal with one and the same objective state of affairs (to make the assumption we would have to assume that both at least refer to the same objective situation. But how can we

assert that 'they both' refer to the same situation when 'they both' never make sense together? Besides, statements about what does and what does not refer can be checked only if the things referred to are described properly, but then our problem arises again with renewed force.) Hence, unless we want to assume that they deal with nothing at all we must admit that they deal with different worlds and that the change (from one world to another) has been brought about by a switch from one theory to another. (*SFS*, p. 70)

For example, it is difficult to deny that, as Kuhn so thoroughly illustrates, the terms *element* and *mixture* have referred to different objects at different times in the history of chemistry. As for the furniture of the world, Feyerabend points out, our epistemic activities have made gods disappear and replaced them with heaps of atoms in empty space. Such talk has riled the many who know in their bones that the world has not changed even though they cannot prove it. For them this relativism is the most obnoxious consequence of the problem of incommensurability.

Some years ago I would have discussed the arguments of Putnam, Scheffler, Shapere, and others who insisted on distinguishing the meaning of a term from its reference as a way to solve this problem of incommensurability.[21] I would have pointed out, as Feyerabend did, that since Bohr's analysis of EPR we know that "there are changes which are not the results of a causal interaction between object and observer but a change of the very conditions that permit us to speak of objects, situations, events" (*SFS*, p. 70). I would have also pointed out, as I did elsewhere, that this "ontological" problem of incommensurability disappears within a sophisticated relativism:

> The issue of whether the world changes when we change basic theory does not arise within the relativity of science. To conceive of the world is to conceive of it within a frame of reference. To ask whether the world really changes when we change frames of reference is comparable to asking (in the Special Theory of Relativity) whether the mass really changes when we change frames of reference. It is just not a sensible question.[22]

The pursuit of these issues, however, does not fit the purposes of this chapter. As it turns out, Feyerabend himself became a critic of relativism in later years. This drastic change in his position, as we will see shortly, simply introduces into his treatment of the issue of reality one of the main themes of his philosophy of science:

> A theory (model, sketch) of a historical process says both too little and too much. It says too little because it starts from a fragment of what it wants to represent. But it also says too much because the fragment is not just described, it is subdivided into essence and accident, the essence is generalised and used to judge the rest.[23]

Relativism proves unsatisfactory, then, because it is a *theory* of a historical process, knowledge. Relativists, Feyerabend believes, accept the assumption that "the theories, facts, and procedures that constitute the (scientific) knowledge of a particular time are the results of specific and highly idiosyncratic historical developments".[24] At the same time they reject the assumption that "what has been found in [an] idiosyncratic and culture-dependent way . . . *exists* independently of the circumstances of its discovery" (ibid., p. 394). They hold, for instance, that "atoms exist given the conceptual framework that projects them" (ibid., p. 403). The trouble with relativism, Feyerabend argues, is that

> traditions not only have no well-defined boundaries, but contain ambiguities and methods of change which enable their members to think and act as if no boundaries existed: potentially every tradition is all traditions. Relativizing existence to a single "conceptual system" that is then closed off from the rest and presented in unambiguous detail mutilates real traditions and creates a chimera. (ibid., pp. 404–405)

This approach echoes Feyerabend's approach to theories of scientific rationality: every one of them, too, circumscribes the practice of science to a single system (of methodological norms or standards), which is then closed off from the rest (in that it is the only acceptable one). But standards, Feyerabend says, "are intellectual measuring instruments; they give us readings not of temperature, or of weight, but of the properties of complex sections of the historical process" (*SFS*, p. 37). To presume that we can apply such standards universally to the practice of science is similar to supposing that we can satisfactorily answer "the question what measuring instruments will help us to explore an as yet unspecified region of the universe. We don't know the region, we cannot say what will work in it . . . (as an example consider the question how to measure the temperature in the center of the Sun, put at about 1820)" (ibid.). Likewise, relativizing existence to a single conceptual system means relativizing it to a single intellectual tradition, and this requires that we ignore how traditions may change, borrow from others, adapt to new circumstances; in short, it means trying to force an abstract "time-slice" on an open-ended historical process.

Realism, at least in its most common varieties, would fare no better in this regard:

> Both objectivism (and the associated idea of truth) and relativism assume limits that are not found in practice and postulate nonsense wherever people are engaged in interesting though occasionally difficult forms of collaboration. Objectivism and relativism are chimeras.[25]

Thus Feyerabend came to believe in the need for a new approach to the problem. He pointed out with the relativists that in auspicious circumstances the entities we project "do indeed 'appear' in a clear and decisive way" (ibid., p. 404). But then he reminded us, with Bohr, that such appearances may be regarded as phenomena "that transcend the dichotomy subjective/objective" (ibid.), which underlies the clash between realism and relativism. "They are 'subjective'", he says, "for they could not exist without the idiosyncratic conceptual and perceptual guidance of some point of view . . . But they are also 'objective': not all ways of thinking have results and not all perceptions are trustworthy" (ibid.). This is so, because the material humans face "offers resistance; some constructions . . . find no point of attack in it and simply collapse" (ibid., p. 405).

In several essays that Feyerabend wrote toward the end of his life, he elaborated the same theme. Quantum theory, one of the best confirmed theories we have, he reminded us, "implies, in a widely accepted interpretation, that properties once regarded as objective depend on the way in which the world is being approached".[26] Nevertheless, he warned, "at this point it is important not to fall into the trap of relativism" (ibid.), for as he had just pointed out, *"nature as described by our scientists is indeed an artifact"*, but an artifact *"built in collaboration with a Being sufficiently complex to mock and, perhaps, punish materialists by responding to them in a crudely materialistic way"* (ibid.).

Nature as described by scientists is not "nature In And For Herself, it is the result of an interaction between two rather unequal partners, tiny men and women on the one side and Majestic Being on the other" (ibid.). This emphasis on such elusive "Being", I believe, marks the return of Feyerabend to some sort of realism, although clearly not to any traditional form, and not quite to an internal realism à la Putnam either. It comes closer, perhaps, to C. A. Hooker's 'evolutionary realism',[27] in that it uses evolutionary motifs and insists on pluralism (a Feyerabend influence on Hooker). Not all interactions, says Feyerabend, produce beneficial results:

> Like unfit mutations, the actors of some exchanges (the members of some cultures) linger for a while and then disappear (different cultures have much in common with different mutations living in different ecological niches). The point is that there is no only one successful culture, there are many and that their success is a matter of empirical record, not of philosophical definitions. (ibid.)

Feyerabend is not limited to pointing out that "the world is much more slippery than is assumed by our rationalists", for "there is also a positive result, namely, an insight into the abundance that surrounds us and that is often concealed by the imposition of simpleminded ideologies" (ibid., p. 99). It is true, he says,

> that allowing abundance to take over would be the end of life and existence as we know it—abundance and chaos are different aspects of one and the same world. We need simplifications (e.g., we need bodies with restricted motions and brains with restricted modes of perception). But there are many such simplifications, not just one and they can be changed to remove the elitism which so far has dominated Western Civilization. (ibid.)

I have argued elsewhere for an evolutionary relativism that offers the flexibility that Feyerabend demands and that fits in well with the open-ended historicity of knowledge that he so ably defends.[28] Thus I do not believe that in order to transcend the common dichotomy realism/relativism it is necessary to make reference to an ineffable Being. Others may disagree with different aspects of Feyerabend's complex view on the issue of relativism (which changed considerably after his sophisticated defense of a practical relativism in *Farewell to Reason*).[29]

But we should all agree that his latest view enlightens us about some important hurdles that await any theory of knowledge. Far from holding some whimsical and (easily?) refutable doctrine, Feyerabend has pushed time and again the frontiers of the philosophy of science.

5. Conclusion

I hope I have been able to convey to the reader some of the complexity and the richness of Paul Feyerabend's epistemology of science, some of the depth of his analysis of the practice of science, and some of the principled ways in which the problems he uncovered, and with which he struggled throughout his career, are consequences of that analysis. Of course, his philosophy went well beyond the issues to which I have alluded in this chapter, but I must content myself with these approximations and simplifications of a body of thought that in itself exemplified the open-ended character of knowledge.

Notes

1. Paul K. Feyerabend, 'Philosophy of Science Versus Scientific Practice: Observations on Mach, His Followers and His Opponents', in his *Problems of Empiricism: Philosophical Papers*, Vol. 2 (Cambridge: Cambridge University Press, 1981), p. 80.

2. This analysis reached its fullest expression in Feyerabend's *Against Method* (London: New Left Books, 1975) (henceforth, AM).

3. This particular point is developed in detail in AM, chap. 6, pp. 69–80. The account of the Tower Argument appears in chaps. 6–8, and the account of the telescope follows it in chaps. 9–11 (chaps. 9–10 in the third edition).

4. See, for example, R. Rorty, *Philosophy and the Mirror of Nature* (Princeton: Princeton University Press, 1979), p. 246.

5. Ibid., p. 101. The original quotation and the ones that follow below are from Galileo Galilei, *Dialogue Concerning the Two Chief World Systems* (Berkeley: University of California Press, 1953) (appropriate references in *Against Method*).

6. In Galileo's time the concern would have been about the transition from the superlunary to the sublunary region.

7. It is wise to take such stories with a grain of salt. Nevertheless, Einstein did write to Born, concerning the latter's remark that Freundlich's analysis of the bending of light near the sun and of the redshift showed that Einstein's formula was not quite right:

> Freundlich . . . does not move me in the slightest. Even if the deflection of light, the perihelial movement or line shift were unknown, the gravitation equations would still be convincing because they avoid the inertial system (the phantom which affects everything but is itself not affected). It is really strange that human beings are normally deaf to the strongest arguments while they are always inclined to overestimate measuring accuracies. (Quoted in AM, p. 57n. Feyerabend's emphasis).

8. P. K. Feyerabend, *Science in a Free Society* (London: New Left Books, 1978), (henceforth, SFS), pp. 142–143.

9. T. S. Kuhn, *The Structure of Scientific Revolutions*, 2nd ed. (Chicago: University of Chicago Press, 1970), (henceforth SSR), p. 78.

10. SSR, pp. 23–24 and 157–158.

11. This account is based principally on Feyerabend's famous essay 'Consolations for the Specialist', which first appeared in I. Lakatos and A. E. Musgrave (eds.), *Criticism and the Growth of Knowledge* (Cambridge: Cambridge University Press, 1970).

12. In P. A. Schilpp (ed.), *Albert Einstein: Philosopher-Scientist*, 3rd ed. (La Salle, Ill.: Open Court, 1982), p. 684.

13. See, for example, *Radical Knowledge*, (Indianapolis: Hackett, 1981), and 'Evolution and Justification', *The Monist*, 71, 1988. In these and other essays I argue for a social conception of scientific rationality.

14. T. S. Kuhn, 'Notes on Lakatos', in R. C. Buck and R. S. Cohen (eds.), *PSA 1970: In Memory of Rudolph Carnap, Boston Studies in the Philosophy of Science*, Vol. 8 (Dordrecht: D. Reidel, 1971), p. 144.

15. See Feyerabend's landmark essay 'Explanation, Reduction and Empiricism', now reprinted in his *Realism, Rationalism and Scientific Method: Philosophical Papers*, Vol. 1, (Cambridge: Cambridge University Press, 1981), pp. 44–96.

16. Often a similar point was made on the basis of the impossibility of translation that incommensurability was supposed to imply. See, for example, H. Putnam, *Reason, Truth and History* (Cambridge: Cambridge University Press, 1981), p. 114.

17. 'Reply to Criticism: Comments on Smart, Sellars and Putnam', reprinted in *Philosophical Papers*, Vol. 1, p. 113.

18. P. W. Bridgman, *The Logic of Modern Physics* (New York: Macmillan, 1927).

19. C. G. Hempel, *Philosophy of Natural Science* (New Jersey: Prentice-Hall, 1966), chap. 7.

20. 'Putnam on Incommensurability', reprinted in his *Farewell to Reason* (London: Verso, 1987), p. 272.

21. H. Putnam, *Meaning and the Moral Sciences* (London: Routledge and Kegan Paul, 1978); I. Scheffler, *Science and Subjectivity* (Indianapolis: Bobbs-Merrill, 1967); D. Shapere, 'Meaning and Scientific Change', in R. G. Colodny (ed.), *Mind and Cosmos* (Pittsburgh: University of Pittsburgh Press, 1966). For an interesting discussion see A. N. Perovich, 'Incommensurability, its Varieties and its Ontological Consequences', in G. Munévar (ed.) *Beyond Reason: Essays on the Philosophy of Paul K. Feyerabend* (Dordrecht: Kluwer, 1991), pp. 313–328.

22. *Radical Knowledge*, pp. 56–57.

23. 'Realism', in C. C. Gould and R. S. Cohen (eds.), *Artifacts, Representations and Social Practices* (Dordrecht: Kluwer, 1994), p. 207.

24. 'Realism and the Historicity of Knowledge', *Journal of Philosophy*, 86, 1989, p. 393.

25. 'Potentially Every Culture is All Cultures', *Common Knowledge*, 3, 1994, p. 20.

26. 'Art as a Product of Nature as a Work of Art', *World Futures*, 40, 1994, p. 98.

27. See his *A Realistic Theory of Science* (New York: State University of New York Press, 1987).

28. See especially my 'Evolution and the Naked Truth', in M. Dascal (ed.), *Cultural Relativism and Philosophy* (Leiden: E. J. Brill, 1991), pp. 177–194.

29. 'Notes on Relativism', in *Farewell to Reason*, pp. 19–89.

7

John Preston

Science as Supermarket

'Postmodern' Themes in Paul Feyerabend's
Later Philosophy of Science

The work Paul Feyerabend published in scattered journals during the 1990s mainly comprises essays and *pièces d'occasion*, many of which are being brought together in the form of a book, *The Conquest of Abundance*.[1] But only the last of his books that he saw into print, *Three Dialogues on Knowledge* (1991), has received the attention it deserved.

Philosophy of science is perhaps the area of philosophy in which 'postmodernism' has had the least penetration and has been least discussed. My intention here is to clarify both Feyerabend's last work and the nature of the postmodern by situating that work relative to three different positions in the philosophy of science which have been *called* 'postmodern'. However, I am less concerned to clinch the case for Feyerabend's having become a postmodernist than to use that position as foil against which accurately to convey and critically evaluate the central themes in his later philosophy.

'Postmodernism' in the Philosophy of Science

I begin with a very mild form of the postmodern outlook.[2] The term has recently been applied, by Jordi Cat, Hasok Chang, and Nancy Cartwright, (Cat, Chang & Cartwright [1991]) to the work of John Dupré. The (overlapping) themes in Dupré's work which they take to license this characterization are as follows:[3]

- His explicit avowal of the 'disunity of science': Science is not, and could never be, a single, unified project (Dupré [1993], p. 1).
- His alignment with what has been dubbed 'the new empiricism', according to which science is "a disjoint collection of models whose range of application is not fully specified and whose effectiveness and accuracy vary considerably within that range" (Rouse [1987], p. 85).
- His identification with the political left.
- His ontological pluralism: "There are countless kinds of things, . . . sub-

ject each to its own characteristic behaviour and interactions" (Dupré [1993], p. 1).

- His opposition to theory-reductionism: although macroscopic objects can be physically smashed into smaller and smaller bits, theories in 'higher-level' sciences are not reducible to those in physics and chemistry. This is because "the individuals that would have to be assumed for the derivation of the macro-theory cannot be identified with those that are the subject matter of the descriptive theories at the next lower level" (Dupré [1983], p. 333).
- His antiessentialism: although there are objective divisions between some distinct kinds of things, questions about the kind(s) that things belong to only make sense relative to a specified goal underlying the intention to classify the objects (Dupré [1993], p. 5).

I do not myself see why this view should be characterized as "postmodernist",[4] but I shall not pursue this.

The Methodological and Theoretical Disunity of Science

We can now note Feyerabend's claim to be postmodern in this very same way. With the possible exception of an explicit political commitment (not to be dealt with here), his later work clearly contains *all* of the above-mentioned themes. Those he endorses most vociferously are clustered around his recognition and approval of the 'disunity of science', which is the central theme in Feyerabend's later work. Not only did he stress it most often and most vigorously, it is also the premise that makes the most sense of almost all the other things he felt it important to say. Naturally, this theme does not come 'out of the blue' into Feyerabend's work. In fact, it is the conclusion he drew from *Against Method*. He makes this absolutely clear in an introduction to the Chinese edition of that book, reprinted in the second and third English-language editions (1988, 1993), where he tells us that the whole book tries to establish the thesis that "*the events, procedures and results that constitute the sciences have no common structure*; there are no elements that occur in every scientific investigation but are missing elsewhere" (AM2, p. 1; cf. [1991b]). This way of framing *Against Method*'s attack on methodological monism suggests that its conclusion, that there is no single scientific method, can be stated in a new way: Feyerabend explicitly tells us that one consequence of the basic thesis of *AM* that he did *not* develop there is that *there can be many different kinds of science* (AM2, p. 3; [1991b]). Western or 'first-world' science, he protests, is but one science among many.

Feyerabend sometimes expressed his disunity of science thesis in the claim that there is no such thing as 'science':

> [T]here is no unified and coherent entity, 'science', which can be said to be successful . . . it is not 'science' that is successful—some so-called sciences are a sorry sight—but particular assumptions, theories, and procedures are. ([1993a], p. 197)

According to Feyerabend, philosophers, scientists, and the educated public are possessed by a certain image of science. The crudest version of this 'classical' image has it that science is 'an undifferentiated monster' ([1991b]). More sophisticated versions, familiar from Newton, Herschel, and Whewell, and taken up by philosophers such as Kant, Mill, Carnap, Hempel, Nagel, and Popper, postulate an identifiable hierarchy of structures from data to theory. Feyerabend assures us that this image of science, a 'harmonious and rather overwhelming fiction' that generates misleading reactions (both positive and negative), is now being dismantled by "a series of developments in the philosophy, the history, and the sociology of science, as well as in the natural and social sciences themselves" ([1995b, pp. 807–808; cf. [1991d], p. 28; [1993a], p. 190). There are many breaks in the supposed hierarchy from fact to theory.

Science, according to Feyerabend, exhibits neither methodological nor theoretical unity. A look at nineteenth-century physics, he suggested, reveals competing perspectives such as the atomic view, the kinetic and mechanical view, the phenomenological view, as well as other viewpoints. Nineteenth-century science was a collection of heterogeneous subjects, some of which had only the most tenuous relation to experiment, others of which were crudely empirical ([1994c], p. 219). The contemporary situation is no different: the unifications achieved by twentieth-century science have not produced the unity the phrase '*the* scientific worldview' suggests.

Antireductionism was, of course, a long-standing theme in Feyerabend's philosophy of science: in 1962 he was already arguing that a formal account of reduction is impossible. In later work he was just as vociferous:

> The idea that 'peripheral' knowledge claims can be reduced to 'more fundamental ones' and, ultimately, to elementary particle physics, which underlies the idea of a coherent body of scientific knowledge is a metaphysical desideratum, not a fact of scientific practice. ([1994a], p. 100, note 13. See also [1989a], p. 402)

We are stuck with a plurality of different theories that so far have resisted the long-sought-for unification. Any 'unity' apparently revealed by studies of science is a product of deception, self-deception, or wishful thinking, not of careful descriptive analysis ([1993a], p. 190).[5] Although unity of science is a regulative ideal strongly favored by philosophers, Feyerabend sees it as playing very little warranted role in scientific practice.

To support these antireductionist views, Feyerabend refers the reader to that *locus classicus* of the 'new empiricism', Nancy Cartwright's *How the Laws of Physics Lie* (Cartwright [1983]). Some of his own occasional claims to be an empiricist may be treated as ironic.[6] But, as we are about to see, empiricism is present in Feyerabend's later work in more than name, and his later critique of science is of a piece with the productions of the 'new empiricists'.

Different Styles of Research

To those who might concede the methodological and theoretical disunity of science, but still protest that there is such a thing as 'the scientific worldview', Feyerabend devoted an entire essay ([1994b]). A close acquaintance with scientific practice, he argued, reveals that science contains very different research-traditions whose perspectives and products cannot be pasted together into a single coherent worldview.

The contrast Feyerabend concentrated on is between empiricism and rationalism.[7] On the one hand, science contains an 'Aristotelian' tradition, whose dictates encourage scientists to keep in close contact with experience, to avoid weak inferences and 'big' cosmological problems, to favor predictions that will be strongly supported or refuted by clear-cut experiments, and not to follow plausible ideas beyond the limits of their plausibility. As followers of this approach, Feyerabend instances scientists like S. E. Luria ([1994a], p. 92; [1994b], pp. 135–136).

On the other hand, there is what we might call a Platonic (or Parmenidean) tradition that takes a more abstract and mathematical approach. It encourages speculation and is ready to accept theories that are related to the facts in an indirect and complex way. Its adherents advise us to follow up plausible ideas to the bitter end and teach that experience (which includes scientific experiments) may be illusory and superficial. Einstein worked in this latter tradition.

When tracing them back to their Greek ancestry, Feyerabend suggests that each of these traditions comes with its own conception of reality. In relinquishing scientific realism, in declaring that "what we need to solve problems is experience, and special pleading" ([1994a], p. 88), Feyerabend chose between these conceptions, favoring Aristotelian empiricism over Parmenidean rationalism. However, he recognized that both traditions have influenced major scientific developments, that neither is dispensable, and that they cannot be reconciled. He drew the conclusion that *"science does not contain one epistemology, it contains many"* (ibid., p. 93):

> Science is not one thing, it is many; and its plurality is not coherent, it is full of conflict. There are empiricists who stick to phenomena and despise flights of fancy, there are researchers who artfully combine abstract ideas and puzzling facts, and there are storytellers who don't give a damn about the details of the evidence. They all have a place in science. ([1992b], p. 6, cf. AM3, p. x; [1994a], p. 97, and [1995a], p. 809)

The wide divergence of scientists, schools, historical periods, and complete sciences, Feyerabend argued, makes it impossible to identify a single scientific worldview. Instead, he endorsed John Ziman's contention that there is no single scientific map of reality, but "many different maps of reality, from a variety of scientific viewpoints" (Ziman [1980], p. 19, quoted in Feyerabend [1994b], p. 141 and [1994c], pp. 219–20).

If this is so, one might well ask why the classical image of science has had such a good run for its money. Feyerabend's answer, with which I am in sympathy, is that the idea of a uniform 'scientific view of the world' may be useful *for motivating scientists*. He compares it to a flag ([1994b], p. 147). One might put this idea in Wittgensteinian terms by saying that the idea of the unity of science constitutes part of the 'prose' of science, part of science's self-image. But for nonscientists, Feyerabend insists, this image is disastrous, since it lends to science an impression of being monolithic, and thereby forces people to divide into two camps: those 'for' science, and those against it. One of the later Feyerabend's main messages, I think, is that there is no question of 'taking sides' for or against science. On each occasion where scientific research is appealed to or called into question, one must examine the individual case on its merits. Simply identifying something as 'scientific' is neither to grant it an award, nor to condemn it. 'Demarcation' is thus rendered de trop.[8]

In sum, these postulated methodological, epistemological, and ideological disunities make it unsurprising that Feyerabend explicitly endorsed Dupré's disunity of science thesis (AM3, p. xi, and [FTR]), and that Dupré nominates Feyerabend as the philosopher whose general perspective on science he most closely agrees with (Dupré [1993], pp. 262–263). The more important question, I think, is how far Feyerabend's views go beyond this mild kind of 'postmodernism'.

In case I am giving the impression that Feyerabend is 'postmodern' only about science, I note that within his later work science is not alone in being disunified. The arts, he points out, are just as good an example of disunity, so is commonsense, and philosophy is even less unified than science (TDK, p. 166; [1995a]). So are cultures (other cultures, at least ([1993a], p. 198)). (The idea that cultures are disunified, we shall see, has important repercussions for Feyerabend's relativism.) Feyerabend detected disunity almost everywhere. The only exception he admitted is our own contemporary Western culture, which many of us like to think of as pluralistic. To Feyerabend, it exhibits cultural monotony. At its most general, Feyerabend put his perspective by saying that *"human activities, though closely related to each other . . . are scattered and diverse"* ([1994a], p. 91).

Parusnikova on Postmodernism in the Philosophy of Science

Zuzana Parusnikova, in an essay that poses the question whether a postmodern philosophy of science is possible (and answers it in the negative), has also outlined something called 'postmodernism' that might at least be *thought* to occur in the philosophy of science. The byword of this view, associated with Jean-François Lyotard, is that our intellectual world is fragmented into a plurality of discourses, without the comfort of any overarching 'metanarrative' that would commensurate them. It involves what Parusnikova calls "the rejection of any higher authority for the legitimation of the rules and goals of scientific performance", and the rejection of "any possibility of defending the universal 'validity' of scientific discourse itself, and its supremacy over other kinds of discourse" (Parusnikova [1992], p. 22). She sums up the result as follows:

> The postmodern world is viewed as fragmented into many isolated worlds; it is a collage, a pastiche of elements randomly grouped in a plurality of local discourses which cannot be unified by any "grand" meta-narrative. . . . For postmodernism, both science and philosophy are just specific discourses among many. Further, science itself is not one homogeneous discourse but rather an empty label for a diversity of research areas and activities. According to the postmodern view, there

exist a plurality of sciences playing their own games and generating their own, local rules for what they do. (ibid., p. 23)

Crucially, this view rejects 'methodological normativism', the philosopher's right to legislate methodological rules and goals for scientists. Such a normative conception of methodology was, I have argued elsewhere,[9] *the* central plank of Feyerabend's early work, and his rejection of it was the most important move in his transition to 'epistemological anarchism' and to an apparently 'historical' philosophy of science. But in fact Feyerabend's later work exhibits *all* these features of 'postmodernism'. Let me point to the other commonalties.

In presenting the disunity of science thesis that we have already encountered, Feyerabend explicitly says, several times, that there is no one thing called 'science'; and even that the word *science* is an empty label:

> terms such as SCIENCE or ART are temporary collecting-bags containing a great variety of products, some excellent, others rotten, all of them characterised by a single label. But collecting-bags and labels do not affect reality; they can be omitted without changing what they are supposed to organise. What remains are events, stories, happenings, results which may be classified in many ways but which are not divided by a lasting and 'objective' dichotomy. ([1994a], p. 93)

Feyerabend's insistence on particularistic case studies instead of broad generalizations is well-known (even though, it has to be said, this is something he preached more than practiced). He made reference to them mainly in order to support his ontological and methodological pluralism. He would certainly have agreed with Parusnikova that studies of science should follow "a genealogical line, as developed by Foucault; they would take the form of case-studies without any grand theoretical backing and without system building" (Parusnikova ibid., p. 30). And he would also agree with postmodernists that philosophy cannot and should not attempt to play a 'foundational' role with respect to other elements of our culture, that is, that philosophy does not offer any special wisdom that other subjects are incapable of supplying.

Feyerabend asked what the alleged universality of science *means*. It had better not mean that scientists in different countries accept the same ideas and use the same methods, because scientists disagree about fundamental matters, even within a single country.

> Nor can the universality of science mean that all scientific laws are universally true and all methods universally applicable. Many laws, methods, disciplines are restricted to special domains. For example, the specific laws of hydrodynamics are neither valid nor thought to be valid in elementary particle physics. ([1992a], pp. 107–108)

The most that the universality of science can mean, he argued, is that science contains universal principles, principles that are supposed to apply to everything (e.g., the second law of thermodynamics). But the universality of a principle does not necessarily mean that it corresponds to universal features of an observer- and history-independent world. Feyerabend, as we shall see, has his own preferred metaphysic to set against that of scientific realism.

Do these views of Feyerabend go beyond the mild 'postmodernism' of Dupré? I think so. As I read Dupré, he is not committed to the dangerous denial of *any* 'metanarrative', a denial that prevents our making sense of conceptual change. I would argue that we do and must, *au fond*, share a conception of human rationality and flourishing that provides the relevant metanarrative. (Note that Feyerabend's earlier epistemological anarchism does not, strictly speaking, rule out such a position. All it rules out is a single distinctive *scientific method or scientific rationality*. But rationality *in general* can survive the nonexistence of scientific method. Although Feyerabend in his better-known and more relativistic phases denied the existence of a single underlying human rationality, he also said plenty of things that can be taken to endorse a 'widened' conception of means-end rationality, a conception that goes well beyond 'scientific rationality'). However, on the present question of whether Feyerabend, in his last work, would have accepted the suggestion that there exists a 'metanarrative', his clear rejection of fixed semantic matrices and of what he called 'objectivism' (see below) situates him with the more radical postmodernists.

Postmodern Semantic Theory

One of the main points of contact between the later Feyerabend and the 'new empiricists' lies in the realization that neither science nor its products should be identified as a set of *linguistic* phenomena. The members of the Kraft Circle, the university philosophy club that included a young Feyerabend in its membership, made two mistakes, he later confessed:

> We assumed that discussing an institution meant discussing its written products. More especially, we assumed that science was a system of statements. Today that seems a slightly ridiculous idea, and the Vienna Circle is blamed for it. But the emphasis on writing is much older. . . . We also assumed, at least initially, that a complicated issue involving major conceptual revisions could be resolved by a single clever argument. (*KT*, pp. 74–75)

The construal of science as a set of complex and multilayered *practices*, rather than as a system of statements, has been taken up most vigorously by Ian Hacking and the whole 'experimental liberation movement' in the philosophy of science (Nancy Cartwright, Peter Galison, Andy Pickering–all of whom Feyerabend refers to admiringly in his later work).[10] It is what Feyerabend referred to when, in his autobiography, he claimed to have become a Wittgensteinian, by virtue of accepting the 'debunking of pure theory' (*KT*, p. 94).

But Feyerabend, I think, goes considerably further down the postmodern road than this. According to a third, and even more recognizably postmodern view, also sketched by Parusnikova, the human predicament involves a pervasive *semantic instability*. The 'deconstructionism' associated with Jacques Derrida develops a 'poststructuralist' view of meaning. In the opinion of its precursor, the structuralist view of de Saussure, meaning is never present in the sign; meaning is always dispersed within a totality of signs and generated by a totality of unstated tactics. Structuralists deny the representational conception of meaning, arguing that meanings are not determined by their referents, but are instead 'caught up' within a relational net of meanings. The meaning of a sign is constituted by what is not present in the sign–by a structure of differences which delimit one sign from other signs. (Parusnikova ibid., p. 32)

This nonrepresentational conception bears comparison with the 'contextual theory of meaning' that Feyerabend deployed in his earlier work, and from which he drew the notorious incommensurability thesis, that certain theories differ from one another so greatly in semantic respects that they cannot be compared with one another (in *these* respects). Although he did *not* think it unavoidable, he argued at that time that this very radical kind of semantic instability is something we should actually strive for, since it gives us the best chance of critically assessing the theories in question. The incommensurability thesis, of course, lingered on in Feyerabend's work long after he had officially eschewed the contextual theory of meaning.

In the mid-1960s, at the very same time that Feyerabend was (disingenuously) presenting himself as never having advocated the contextual theory, Derrida was going beyond the structuralist conception, concluding that "these structures of differences can never be determined and, therefore, meaning can never be completed; *the meaning of a sign is determined by a trace of that other which is forever absent*" (Parusnikova ibid., p. 32).

Insofar as he mentions meaning in his later work, Feyerabend stresses the existence and omnipresence of ambiguity. Concepts, he tells us, are ill-defined, unstable, and ambiguous ([1991a], p. 515). For Feyerabend, just as for the postmodernist, meaning is "fundamentally elusive and slippery" (Parusnikova ibid., p. 21). These features of meaning-relations make meaning a quarry that will always elude philosophers, although this does not imply that *science* will suffer: scientists systematically *ignore* the meanings of their pronouncements, or warp them to fit their own momentary needs. During this phase, Feyerabend disavows the project of giving *any* theory of meaning. He arrives at a semantic nihilism[11] according to which meaning simply *doesn't matter* to science.

Nevertheless, even here, Feyerabend managed to say quite a lot about meaning without having a theory of it. His leitmotif is that texts lack the uniformity that would be required for them to constitute closed semantic domains. The regularities they do contain are often violated, and these violations do not constitute nonsense. Their perpetrators do not leave the domain of meaningful discourse, but merely enter new territory. "[I]ngenious individuals often give sensible answers to allegedly inexpressible questions", and there exist "people who can explain events that baffle or enrage others, failing to recognise the limits of sense postulated by scholars" ([1994d], p. 17). Stable linguistic 'habits' are not necessarily boundaries of sense ([1994c], p. 208). At his most radical, Feyerabend claims that we do not know what even a simple generalization like 'All ravens are black' *means*, since we do not know in advance the conditions under which we would give it up.[12]

Unless we perform "further research" we cannot say whether bizarre cases will count as counterexamples to the generalization or not. Hence the statement "is not a well-defined semantic entity" ([1989b], p. 189). A theory, likewise, instead of being "a well-defined entity that says exactly what difficulties will make it disappear" is revealed as "a vague promise whose meaning is constantly being changed and refined by the difficulties one decides to accept" (*TDK*, p. 41). The tendency of logicians and philosophers to interpret statements in a simple-minded way, and subsequently to refute them, "would have killed science long ago", since "[e]very scientific theory, interpreted in a literal way, is in conflict with numerous facts!" (ibid.). In sum, "we have no reason to assume that our ways of conveying meaning have *any* limits" ([1994d], p. 20, emphasis added).

Because neither statements nor theories have a determinate meaning, both the sort of legislation that 'linguistic philosophers' (allegedly) indulge in *and* the semantic thesis of incommensurability are impugned as restrictive. Those who, following Kuhn, have claimed that the transition from one comprehensive theory to another involves an act of conversion similar to a gestalt-switch are indulging in a priori history of science to the same extent as the logical positivists they superseded. If we look at history, we do not find that scientists on either side of paradigm-transitions talk past one another ([1993b], p. 7). In particular, says Feyerabend, the transition from classical physics to the quantum theory was not of this kind. The 'conversion philosophy' is therefore inadequate as an account of science; and it fares no better when applied to cultures (ibid., p. 8).

These semantic views go beyond the milder forms of postmodernism in holding that elucidation of a language-game, even 'locally', is strictly impossible. Not only is there no 'permanent neutral matrix' (or 'metanarrative') within which inquiry takes place, but there are no *semantic* matrices within which meaning can be determined. I personally have no love for the conception of meaning that is operating here, since I think its main theses (that meaning is an activity, that it is never determined, etc.) are either literally unintelligible or plainly false.[13] It exaggerates semantic instability, ambiguity, imprecision, and our alleged inability either to grasp or to specify meaning. Intent on emphasizing that there can be changes in whether a sentence counts as meaningful, as well as in the meaning of a given sentence, Feyerabend accommodates these phenomena only by denying that there are ever any bounds of sense. His only replacement for this idea is the notion that some possible combinations of words constitute 'stable linguistic habits', and others get uttered infrequently, if at all. Finding no fixed and final limits to sense, Feyerabend supposes that, when it comes to language, anything goes. I consider this to be a step *backward* from his (already inadequate) contextual theory. But I think it is clear that Feyerabend and Derrida were thinking along roughly similar lines in this respect.

The Postmodern Philosopher of Science?

What role, on this sort of view, is left for the philosopher? Perhaps only that which Richard Rorty suggests, the provision of 'edifying' and ironic commentaries on (what used to be called) 'first-order' activities. Derrida has been thought to supply a less liberal and less compromised version of the same ideal, according to which the philosopher's function is to question, annoy, tease, disturb, destabilize, provoke, and poke fun at the activities of those he or she is concerned with. The aim of such philosophical activity is not a positive one like knowledge, understanding, or conceptual clarity, but rather a *destructive* one. On such a conception, as Parusnikova puts it:

> [W]e do not try to achieve the maximum possible clarity of understanding, but rather criticise and subvert prevailing views, deliberately 'misunderstand' and 'misread' texts. We do not appeal to any general authority; we just play games whose rules can be changed when we become bored. The message is: do not take anything too seriously, it is not worth the trouble. (Parusnikova ibid., p. 25)

The postmodern philosopher has no interest in 'elucidating' science: "scientific discourses are autonomous; they may overlap, but that is the business of the participants in the discourses themselves. No 'advisers' from outside are let in" (ibid., p. 28). Feyerabend's later work, it seems to me,[14] fits this description rather well. He was *never* interested in 'clarifying' or merely explaining the activities of first-order agents. In his earlier work, his stated aim was that of making scientific progress *within* philosophy. In his later incarnation, as is well-known, this scientistic conception of philosophy is buried (although it never disappears absolutely), and it is buried under something like the postmodern view. (The most obvious point of difference, as we shall see, is on the issue of autonomy). The worst philosophical crime Feyerabend could conceive was to take on what Nietzsche called 'the spirit of gravity' (which is what he discerns in Popper's 'rationalism').

Unfortunately, Feyerabend does not supply any resolution to the puzzle that Parusnikova invokes, that of the relation the postmodernist philosopher can have to scientific discourses. She worries that the 'games' scientists play are "so highly specialised and inaccessible to outsiders that a postmodern philosopher (unless she is a trained scientist) could simply never communicate with scientists" (Parusnikova ibid., p. 26). Although Feyerabend thought science could and should be assessed by lay people (in 'democratic councils') he himself *was* trained as a scientist, and he despised those philosophers of science who know nothing of science 'from the inside'. But there is a strong tension between his social populism and the postmodern denial of the unity of science thesis. If scientific discourses are, and must remain, so highly specialized and diffuse, what becomes of Feyerabend's demand that they be evaluated by laypeople? One of the central political points in Neurath's advocacy of the unity of science thesis resurfaces here.[15]

Against Scientific Realism

Feyerabend's more radical version of postmodernism has antirealist implications, bearing negatively both on realism about scientific *theories*, and on realism about scientific *entities*.

One of the major arguments in favor of scientific realism deploys the idea of the success of science. Feyerabend's later work, so far largely undiscussed by realists, constitutes a sustained critique of the meaning and implications of this slogan. His main reservations about the realist argument are as follows.

First, he challenges the idea that science can be credited with success, and this challenge naturally begins with the recognition that there is no such thing as 'science'. We have already noted his protest that only particular assumptions, theories, and procedures, not science itself, can be deemed successful ([1993a], p. 197). That there are such successes, he insists, *"is a fact, not a philosophical position"* ([1994b], p. 135; [1994c], p. 217). But many theories that have surprising successes in one area fail in others ([1993a], pp. 191–2). Newtonian mechanics is an example. This makes it impossible to produce a single success-or-failure verdict on a theory. Second, the disunity of science means that its ingredients have very unequal strength, failure is more common than success, and success is the result of methodological daring, not of adherence to a clear rationality or uniform procedures (*AM2*, p. 2; [1991b]): "Successful research does not obey general standards; it relies now on one trick, now on another; the moves that advance it and the standards that define what counts as an advance are not always known to the movers" (*AM2*, p. 1).

One consequence of the disunity of science thesis, Feyerabend tells us, is that *scientific successes cannot be explained in a simple way* (*AM2*, p. 1). We cannot give a recipe for discoveries. "All we can do is to give a historical account of the details, including social circumstances, accidents, and personal idiosyncrasies" (ibid., p. 2). This is exactly the kind of localist case-study of which postmodernists approve.

Another consequence is that

> the success of 'science' cannot be used as an argument for treating as yet unsolved problems in a standardized way. That could be done only if there are procedures that can be detached from particular research situations and whose presence guarantees success. The thesis says that there are no such procedures. (ibid.)

And a further implication of the fact that science is a collection of different approaches, some successful, others wildly speculative, is that "there is no reason why one should disregard what happens outside of it" ([1994c], pp. 220–221), "'Non-scientific' procedures cannot be pushed aside" (*AM2*, p. 2). Feyerabend's third major reservation about the argument from the success of science is therefore that insofar as science can be said to be successful, *plenty of nonscientific activities are successful too* (see [1992b], p. 8; [1994a], p. 98).

Of course, this raises the question "In what way(s), in terms of what standard(s) can the success of activities and cultures be gauged?" Feyerabend is often interpreted as holding that comparison of unlikes is out of the question. And perhaps this is what he means when he says that "[t]here is no objective principle that could direct us away from the supermarket 'religion' or the supermarket 'art' towards the more modern, and much more expensive supermarket 'science'" ([1994b], p. 146). When discussing the comparison of cultures, he tacitly admits that the success of nonscientific activities and cultures is not measurable in terms of all the virtues that characterize the best scientific theories (predictive accuracy, notably). Nevertheless, he does not shrink from the comparison: he proposes to gauge the success of cultures in terms of general humanitarian criteria such as whether they sustain lives that are meaningful and desirable to the people concerned. (More pertinently: whether they did so *before* they were ruined by the onslaught of Western civilization (*KT*, p. 143)). These are the only criteria by which *both* scientific and nonscientific activities can be assessed (which precludes our being relativists about value-judgments). If we are intent on comparing unlikes, Feyerabend insists that many (all?) traditions and cultures, some of them wildly unscientific, succeed in the sense that they enable their participants to live moderately rich and fulfilling lives ([1994c], pp. 220–221). Because nonscientific activities measure up just as well in terms of these virtues, we should not discriminate against them by granting science a special status.

Feyerabend's attack on the argument from the success of science is worth taking seriously, if only to force specification of the respects in which components of science are successful. I concur with those who find his refusal fully to admit the technological and instrumental fruits of scientific activity disingenuous. Insofar as science can be said to be a single 'way of dealing with nature', it is unsurpassed. But, of course, part of his point is that science is *not* a single such way.

Feyerabend's suggestion that nonscientific cultures are successful too misses the point. Technological-instrumental success is part and parcel of the conception of good functioning that all humans share. As such, it is not something one can be indifferent to. One may have all sorts of reservations about the uses to which science and technology are put, and the aims which they are sometimes made to subserve but, where scientifically attainable goals can be specified and agreed on, scientific means often constitute our most efficient ways of attaining them.

Beyond Objectivism and Relativism

In his last works, Feyerabend, once designated (by Lakatos) 'our most brilliant cultural relativist', apparently recanted relativism, coming to characterize it as a 'trap' ([1992d], p. 8; [1994a], p. 98). This emphatically does not mean that he returned to what he called 'objectivism', the view that truth is not relative to culture. He admitted to having been a relativist (*TDK*, p. 157), although insisting that even in his most extravagantly relativistic moods, he never endorsed the idea that any standpoint is as good as any other ([1991a], p. 503). He never doubted that the 'practical intentions of the relativist' must always be protection and tolerance (*TDK*, p. 42), and he never addressed the obvious objections to the idea that relativists are in a position to promote these intentions.

Feyerabend's retreat from relativism proceeded as follows. Through much of the 1980s, he had defended what he thought of as Protagorean relativism (see [1984], *FTR*, and Preston [1995]). But by the end of that decade, Feyerabend had come to think of relativism as "a very useful and, above all, humane approximation to a better view", albeit one which he admitted to not yet having found (*TDK*, p. 157). The first of his *Three Dialogues on Knowledge*, dated 1990, still contains expressions of Protagorean relativism (pp. 19, 26–41), but they appear alongside inklings of his impending retreat (pp. 16ff. Cf. [1989a], section 4). In an essay published in 1991, replying to critics of his work, Feyerabend points out that the term 'relativism' is ambiguous, and that he is a relativist only in some senses ([1991a], p. 507. Cf. *TDK*, p. 150). He also confessed to having had a change of heart since *Farewell to Reason* (1987), and the first thing he recanted was 'democratic relativism', the normative ethical view that all traditions should be given equal rights and equal access to power. He became uneasy with this view because it reflects on traditions "from afar" ([1991a], pp. 508, 511); it is an 'armchair' relativism (*KT*, pp. 151–152). More significantly, he regretted having introduced the *general* principle 'hands off traditions' because not all cultures are isolated, as relativists represent them ([1991a], p. 509; *TDK*, pp. 16ff). He therefore replaced these ideas with the suggestion that traditions have intrinsic value ([1991a], p. 510).

These reservations about relativism were amplified over the last few years of Feyerabend's life. Their crux is the idea that neither views nor cultures are closed domains. Let us first note the connection with the postmodern semantic doctrines we already unearthed. Feyerabend lost confidence in our ability to identify any unambiguous basic principles underlying a culture. The 'conversion philosophy' beloved of those infatuated with incommensurability implies that each text or theory constitutes a closed semantic domain, a 'field' that gives meaning to the terms it contains. But since, as Feyerabend recognized, there is communication during revolutionary transitions, the idea that only 'systems' give meaning to what is said is inadequate (*TDK*, p. 18). Relativism *would be correct* if scientific nature and scientific procedures were well-defined, unambiguous, and 'closed'. But they are not ([1992b], p. 8; see also [1994a], p. 98; [1994d], p. 17, and *KT*, p. 151). We are a long way from the disaster systematized by champions of incommensurability ([1994d], p. 19). Texts do not constitute closed or unified semantic domains, with clear limits between what lies within and what lies beyond the bounds of sense.

Forms of relativism according to which whatever one says is valid only 'within a certain system' wrongly presuppose that all the elements of a system are unambiguous (*TDK*, p. 151). Likewise, cultural relativism wrongly presupposes that cultures have 'systems of thought' (ibid., pp. 17, 42) and that one cannot learn new ways of life. These assumptions are very unrealistic. For if one *can* learn a new way of life, "one system is potentially all systems, and the restriction 'relative to system A', *while useful for special purposes, loses its power as a general characterization of knowledge*" (ibid., p. 152). The result is that relativism "correctly describes the relations between the frozen systems or prisons which some people make for themselves. It is a good account of the ideas of people who dislike change and turn difficulties of communication into matters of principle" (*TDK*, p. 151). Those who are unwilling to change and who are afraid of changing others (those who are determined to 'protect non-Western cultures', for example) will therefore find themselves in an artificial world that is perfectly described by the 'philosophical principles of incommensurability and indeterminacy of translation' ([1993c], p. 9).[16] But intercultural understanding cannot exist without contact, and contact *changes* the parties engaged in it:

> Relativism . . . believes that it can deal with cultures on the basis of philosophical fiat: define a suitable context (form of life) with criteria etc., of its own and anything that happens in this context can be made to confirm it. As opposed to this, real cultures change when attempting to solve major problems and not all of them survive attempts at stabilisation. The 'principles' of real cultures are therefore ambiguous, and there is good sense in saying that every culture can in principle be any culture. ([1994a], p. 98)

I think this is a significant step *beyond* the second kind of postmodernism. The perpetual change of cultures consequent upon their interaction belies the postmodern suggestion that discourses are *autonomous* (see the quote from Parusnikova, p. 28, above).

It is ironic that these considerations that made Feyerabend turn his back on relativism are among those which Ernest Gellner (one of his erstwhile bêtes noires) had urged against Peter Winch's relativistic philosophy of social science in the mid-1960s.[17] They are what led Feyerabend to one of his last slogans, that 'both objectivism and relativism are chimeras' ([1994d], p. 20; see also [1991a], pp. 503, 507–508, 513–515; *KT* p. 152, and [*FTR*]). Objectivism and relativism, Feyerabend came to feel, are 'cantankerous twins', with a common ancestry in Plato's mistaken semantic view that concepts are stable and clear ([1991a], pp. 515–516). In a recently printed new preface to *Farewell to Reason* he claims that his views should not be classified as relativist, since this

> assumes that the plurality I recommend consists of clearly separated domains, each held together by clear rules, criteria and ways of seeing things. A relativism so conceived is simply objectivism broken into pieces and multiplied. The pieces rule over [their] members in the same tyrannical way in which objectivism tries to rule over everybody. But cultures and ways of life lack the stability and the exclusiveness of relativistic domains. (*FTR* [New preface])

Feyerabend came to believe that scientists (and others engaged in 'first-order' activities) must simply learn to live with ambiguity and contradiction, and that it is only philosophers (those seeking a 'higher-order' account of first-order activities) who insist that these accounts must be clear, unambiguous, and contradiction-free. This, he says, "is the reason why the presentation of scientific *results* differs so drastically from what happens during *research*, i.e., while people are still *thinking*, and gives such a misleading picture of it" ([1994d], p. 21). Scientific 'facts' and 'principles' are, according to Feyerabend, artifacts produced not by the scientists in the 'core set' who deal with the discipline's experimental problems, but by *outsiders* who disseminate the results of the core set's efforts:

> Scientists of the same field but outside the core set . . . read the verdict in a more uniform and slightly dogmatic manner. Neighbouring disciplines soon speak of 'facts' and 'principles'. Philosophers of science, whose distance from core sets can be enormous, then support these facts by showing the facts' rationality, while popularisers, textbooks, introductory lectures regard the facts as proof of the Forward March of Science. ([1991c], p. 29)

This sort of account, chiming with Peter Medawar's suggestion that a scientific paper is something of a fraud,[18] strikes a chord with working scientists and connects up with Feyerabend's own metaphysic. If we take the attitude Feyerabend recommends, the philosophy of science is supposed to wither away, to be replaced by history and sciences that can take care of themselves.

Feyerabend's New Metaphysic

Because the scientific realism of Feyerabend's earlier work was underpinned by his normative conception of epistemology, the demise of this conception eventually issued in a retraction of scientific realism.[19] One of the most remarkable things about his last work is that he commits himself, albeit tentatively, to a new metaphysical picture of the world, a clear rival to the picture of mind-independent reality that undergirds scientific realism. The replacement is a metaphysic best characterized as 'social-constructivist'.

Sometimes Feyerabend takes Pickering, Galison, Rudwick, and others he identifies as 'social constructivists' to have shown merely that "scientific facts are *constituted by* debate and compromise, that they harden with the distance from their origin, that they are *manufactured* rather than read off nature, and that the activities that produce and/or identify them form complex and . . . relatively self-contained cultures" ([1995b], p. 808). However, we will not waste time challenging the intelligibility of speaking of facts being constituted by or manufactured in the course of social processes, for Feyerabend's constructivism turns out to be even more radical. He begins his 1992 essay 'Nature as a Work of Art', by declaring that *"nature as described by our scientists is a work of art that is constantly being enlarged and rebuilt by them"* ([1992b], p. 3; [1994a], p. 94). In other words, he says, "[O]ur entire universe . . . is an *artefact* constructed by generations of *scientist-artisans* from a partly yielding, partly resisting material of unknown properties" ([1992b], pp. 3, 8; [1994a], pp. 88, 94; see also [1989a], pp. 404–406). Even though this thesis is tempered, in one essay, with its obverse (that artworks are products of nature) it sounds, like any other kind of social constructivism about physical things, exciting but absurd. Although Feyerabend, unlike advocates of the most extreme possible constructivist position, admits that there is some (albeit weak) 'resistance' to our epistemic activities, the resistance is offered not by particular objects, processes, or properties, but simply by reality, nature, or 'Being' (as he prefers to call it) itself (*TDK*, pp. 152–153). And whatever laws govern this resistance are absolutely unknowable. The resulting view is that our world has a thoroughly Kantian structure: everything we are tempted to think of as a constituent of the world is something created by our own epistemic activities; only 'Being' exists independently of us. But Being is a completely unknowable 'thing-in-itself', a something of which nothing can be said. Feyerabend does not shrink from the idealist implications of this metaphysics. In answer to the obvious objection that constituents of the universe existed long before humans, and therefore could not have been created by them, he insists that scientists, like Platonist mathematicians, misjudge the implications of their own activity in assuming that their subject-matter is independent of the human race.

Surely this does not constitute an adequate reply? As the unsuccessful track-record of phenomenalism shows, our conception of the constituents of the universe is a conception of things which *are* mind-independent. Surely "[a]toms and stars and galaxies exist no matter what anybody thinks about them. They are *real*, not artefacts" ([1992b], p. 4). Feyerabend proposes to undermine what he calls this 'plausible platitude' by inviting us to consider certain aspects of the rise of modern science. Leading Western scientists, in preferring experiment to observation, and mathematical formalisms to ordinary language, replaced natural processes by artifacts. So "modern science uses artefacts, not nature-as-it-is" (ibid., p. 6). We can infer from this that "nature-as-described-by-our-scientists is also an artefact, that nonscientific artisans might give us a different nature and that we therefore *have a choice*" (ibid.) because the resistance nature puts up to our investigations is *not* uniform. Only the misguided assumption that there is a uniform scientific method or a unified scientific world-picture leads us to think that nature has what we might call an *inherent structure*, which cannot be changed by human intellectual activities. And this assumption, of course, Feyerabend thinks he has disposed of by his insistence on the sheer plurality of science.

It has to be admitted that Feyerabend's argument against the plausible realist platitude and its associated metaphysic is weak, for it conflates the scientist's representational machinery with what it is supposed to represent. Only Platonists balk at the suggestion that we construct our representational resources (including experiments, formalisms, languages and concepts). But it is hardly necessary to point out that neither human bodies, nor human minds, nor the social processes in which people participate have the kinds of causal powers necessary to *construct* (e.g.) galaxies. Any account according to which such a thing is possible would be flatly incompatible with the science which it claims to be an account of. It is hard to see how even a plural, disunified worldview could tolerate such an obvious and easily identifiable contradiction. In using the artifacts characteristic of scientific conceptual schemes, the scientists Feyerabend refers to were not rejecting the existence of 'natural processes'. In fact, use of the artifacts in question does not make sense *unless* the relevant natural processes exist. If the constructivists' claim is, rather, that humans have the power to construct things *as* galaxies (etc.), we should need clarification of the meaning of this ungrammatical expression.

This is where constructivism seems at its weakest, and the realist's advantage is clear. The realist begins by using the postulate of a determinately structured reality to explain those occasions on which scientists agree. The constructivist then proposes to replace this postulate with the idea that scientists agree when their social situations are structured similarly. This is already unacceptable, since it implies that had the scientists in question not been similarly situated, they would not have converged on the same conclusion. Further, it does not begin to explain those occasions on which scientists in different laboratories, at different times, and under very different social circumstances, manage to reproduce results that lead them to the same conclusions. If the 'resistance' the world puts up is not strong enough to make enquiries converge on a single outcome, what *can* explain the occasions on which such 'closure' is achieved despite the fact that the enquiries have been conducted from within different forms of life?

The constructivist is in an even worse position than a nonspecific antirealist here, for the constructivist explanation of closure in socially similar situations actually precludes there being closure in socially distinct ones. On the constructivist view, closure in situations where scientists are differently situated is not just a miracle, but an impossibility. Its existence would refute the constructivist account. Feyerabend seems too preoccupied with potential scientific *disagreements* to notice this.

The linchpin of Feyerabend's new picture is the idea that Being reacts to different enquiries in different ways:

> [H]aving approached the world or, to use a more general term, Being, with concepts, instruments, interpretations that were the often highly accidental outcome of complex, idiosyncratic, and rather opaque historical developments, Western scientists and their philosophical . . . supporters got a finely structured response containing quarks, leptons, spacetime frames, and so on. The evidence leaves it open if the response is the way in which Being reacted to the approach, so that it reflects both Being and the approach, or if it belongs to Being independently of any approach. Realism assumes the latter; it assumes that a particular phenomenon—the modern scientific universe and the evidence for it—can be cut from the development that led up to it and can be presented as the true and history-independent nature of Being. The assumption is very implausible, to say the least. For are we really to believe that people who were not guided by a scientific world view but who still managed to survive and to live moderately happy and fulfilling lives were the victims of an *illusion*? ([1992a], p. 109, emphasis added. See also *TDK*, pp. 42B43, 152–153)

Instead of presenting others thus, Feyerabend argues, we ought to consider the possibility that Being is more yielding, dynamic, multifaceted, and responsive than contemporary materialists and realists concede (see, for example, [1988], p. 178; [1989a], p. 406). Perhaps

> the gods, saints, demons, souls, and the centrally-structured spaces that played such an important role in [the lives of nonscientific people were] the way in which Being received their approach, so that given this approach they were as real as elementary particles are supposed to be today? ([1992a], p. 110)

So instead of thinking that the procedures of a research program reveal how nature is independently of the interference, we are to think of them as revealing how nature responds to them. The conclusion we have already seen Feyerabend draw, that science contains a plurality of different approaches, is his first argument for this new metaphysics. Because most of these conflicting approaches had results, we cannot legitimately present nature as having (only) the particular complexion ascribed to it by our own favorite theory (see [1992b], p. 7; [1994a], p. 97).

We have also seen a second, closely related reason why Feyerabend proposes that reality should be thought of in this strange new way: recent discoveries in anthropology, the history of astronomy, medicine, mathematics, technology, theology, and other fields show that "non-scientific cultures provided acceptable lives for their members and that the imposition of Western ideas and practices often disturbed the delicate balance with nature they had achieved" ([1994a], p. 98).

Feyerabend's third and final argument for his preferred metaphysic is simply that the quantum theory, one of our best-confirmed theories, implies ("in a widely accepted interpretation") that properties once regarded as objective depend on the way in which the world is approached (ibid., p. 98; [1991b]; [1992d], p. 8).

Some initial problems with social constructivist metaphysics have already raised their heads. Perhaps none of them would have impressed Feyerabend. But one consideration that should have weighed heavily is that his move to constructivism simply does not solve the problem which gave rise to it, the problem which he identified as plaguing relativism. For the ways in which 'Being' responds to inquiry are partitioned *no more distinctly* than the relativist's 'systems of thought'. This social constructivism does not address the facts that theories are open to change and subject to different interpretations. Only the most extreme reductionists would seek to deny that the world is a complex and many-sided thing. But this idea needs more careful handling than Feyerabend devotes to it. To say that the world has a multifaceted nature is one thing; but it is not to endorse Feyerabend's supposition that a completely unknowable entity, 'Being', has the (quite remarkable but logically inexplicable) capacity to respond in different ways to different inquirers, depending on their psychological and social situations. Neither is it to accept Feyerabend's supposition that any ontology which sustains a thriving culture *eo ipso* exists. A world (hyper-world?) crowded with all the things everyone has ever believed in is a world none of its inhabitants can begin to explain.

A New Attitude Toward Cultures?

For Feyerabend, probably the most important implication of his later slogans was what he called a "new attitude towards cultures" ([1994d], p. 22), his alternative to the unrestricted and noncommittal tolerance he associated with relativism, and the intolerance he associated with objectivism. The final major theme sounded in his writings develops from his earlier idea that "one system is potentially all systems" (*TDK*, p. 152; [1989a], p. 404), and comes to be encapsulated in his slogan "potentially every culture is all cultures" ([1994a], p. 98; *KT* p. 152; and [*FTR*]):

> [P]ractices that seem legitimate when referred to a closed framework cease to be sacrosanct. If every culture is potentially all cultures, then cultural differences lose their ineffability and become *special and changeable manifestations of a common human nature*. Authentic murder, torture, and suppression become ordinary murder, torture, and suppression, *and should be treated as such.* . . . Efforts to achieve peace need no longer respect some alleged cultural integrity that often is nothing but the rule of one or another tyrant. ([1994d], p. 22)

In the present postcolonial world, the notion of an authentic culture as an autonomous internally coherent universe is no longer tenable (ibid.). But we cannot return to the modern (that is, prepostmodern) 'objectivist' idea of universally applicable rules and judgments. Having discarded both objectivity and cultural separation, and having emphasized intercultural interactions, those who perceive medical, nutritional, and environmental problems or problems of human rights have to start such processes on the spot and *with due attention to the opinions of the locals* (ibid.):

> [G]eneral talk about the 'objective validity of science' simply misses the point. To me, assertions such as "science teaches us about reality" or "reality is relative to a culture" are metaphysical romances of doubtful merit. They have to be replaced by a piecemeal approach which relies on local measures and whose generalities arise, like the generalities of science, from the experiences that were gained in their course. ([1993a], pp. 199–200)

The 'particularizing' attitude to the study of science is thus complemented, in Feyerabend's last work, by a particularism about cultures. This is undoubtedly one of the insights he drew from his last wife, Grazia Borrini, who has been involved with conservation, as well as health, hunger, and development aid.[20]

The idea that there is a common underlying human nature looks like an earlier version of empiricism: Humean naturalism. Postmodernists, by contrast, are usually taken to *deny* that human nature remains constant across time and culture. The Humean thesis raises questions that cannot be addressed here: is Feyerabend's position an advance on postmodernism? If human *nature* is constant across cultures, can human *rationality* really vary as much as he seems to think? Or are the commonalty of human nature and the potential commonalty of human culture reasons to think that Feyerabend would accept a commensurating 'meta-narrative'? Can his position be assimilated to that adopted by certain Wittgensteinian thinkers, according to whom certain *life-experiences* (rather than principles of rationality) are what people from widely differing cultures have in common?

Conclusion

I have expressed some initial criticisms of the postmodern semantic perspective, the social constructivist metaphysics, and the unsuccessful critique of 'objectivism' toward which Feyerabend gravitated. But I want to end by drawing attention to the more promising themes that emerge from his last work. That work, I would argue, is important for the following reasons. It offers a sustained presentation of the disunity of science thesis, within which I have delineated methodological, theoretical, ontological, perspectival, and epistemological strands (there may be others!). Its critique of the realist's argument from the success of science deserves to be taken more seriously than does Feyerabend's own social-constructivist alternative to realism. Its alignment with antirealism, in particular with Aristotelian empiricism, as well as its other points of contact with the themes of the 'new empiricists', contain the important recognition that science is a collection of practices, not just of propositions. Feyerabend showed an admirable willingness to question the 'conversion philosophy' and the relativism with which his name now seems indelibly associated. It is important to see *through* the (somewhat 'postmodern') way in which he practiced philosophy of science, as well as his overwhelmingly negative account of its role, to his refusal to take sides either for or against science. In this respect, perhaps he *does* deserve to be thought of as Wittgensteinian, for Paul Feyerabend was ultimately a citizen of no community of ideas.

Notes

I am grateful to Steve Fuller and George Reisch for comments on drafts of this chapter.

1. Scheduled to be published soon by the University of Chicago Press.
2. Not the mildest form, though. Nancey Murphy (Murphy [1990]) outlines a sense of the term in which Quine is a paradigm postmodernist, and there has been some discussion of whether *Popper* counts as such!
3. See also my review of his book, Preston [1994].
4. Dupré prefers to think of himself as a 'critical modernist'.
5. At certain points (*TDK*, p. 146; [1989a], p. 401), Feyerabend apparently opposed even *ontological* reduction, on the grounds that in the reduction of molecules to elementary particles, information of one kind is lost and replaced by another kind (the Bohmian 'wholeness' involved in elementary particle processes).
6. Although I suspect he took his own claim to look at history 'as an empiricist' ([1991a], p. 515) seriously.
7. See, for example, [1988], Section 5; *TDK*, p. 138; [1992a], p. 112; [1993c]; [1994b]; and [1995b]. This contrast lies at the heart of Feyerabend's account of 'the rise of Western rationalism', a major theme in his later work, but one to be covered on another occasion.
8. George Reisch [1998] has objected that Dupré's disunity thesis precludes him from demarcating science from nonscience, thus opening the door to 'creation science', and other abominations. Whether or not this cuts ice against Dupré, Feyerabend might accept the conclusion with impunity.
9. Preston [1997a], chap. 1.
10. Feyerabend's positive references to the main protagonists are as follows: to Cartwright: [1991a], p. 505, [1993a], p. 190, [1994a], p. 100, note 13, [1995b], p. 808; to Galison: [1989a], p. 394, *TDK*, p. 141, [1991a], p. 505, [AM3], p. x; [1995b], p. 808; to Hacking: [1991a], p. 505, [AM3], p. xi; [1993a], p. 190; to Pickering: [1989a], p. 394, [AM3], pp. x–xi; [1995b], p. 808. Fine, Latour, Primas, Rudwick, van Fraassen, and unspecified sociologists of science also get occasional mentions.
11. For Feyerabend's commitment to semantic nihilism, see Preston [1995]. I don't mean to imply that Feyerabend admired Derrida: it seems that he didn't (see *AM3*, p. xiv and *KT*, p. 180).
12. Note the falsificationist conception of meaning (which not even Popper subscribed to) inherent in this complaint.
13. See Preston [1995].
14. Here I simply endorse and develop Parusnikova's (p. 25) suggestion.
15. For this suggestion, I am grateful to George Reisch.
16. Note that Feyerabend never understood Quine's indeterminacy thesis, assimilating it to his own thesis of incommensurability.
17. See Gellner [1968].

18. Medawar [1963], occasionally mentioned by Feyerabend.
19. As I have detailed in Preston [1997b].
20. For references to her influence, see especially *FTR*, p. 4 note, p. 318, and *KT*, chap. 14, pp. 175–176.

References

N. Cartwright (1983) *How the Laws of Physics Lie*. Oxford: Clarendon Press.
J. Cat, H. Chang, and N. Cartwright (1991) 'Otto Neurath: Unification as the Way to Socialism', in J. Mittelstrass (ed.), *Einheit der Wissenschaften*. Berlin: Akademie der Wissenschaften.
J. Dupré (1983) 'The Disunity of Science', *Mind*, 92.
J. Dupré (1993) *The Disorder of Things: Metaphysical Foundations of the Disunity of Science*. Cambridge, Mass.: Harvard University Press.
P. K. Feyerabend (1975) *Against Method*. London: Verso. (AM).
P. K. Feyerabend (1984) 'The Lessing Effect in the Philosophy of Science: Comments on Some of My Critics', *New Ideas in Psychology*, 2.
P. K. Feyerabend (1987) *Farewell to Reason*. London: Verso. (FTR).
P. K. Feyerabend (1988) 'Knowledge and the Role of Theories', *Philosophy of the Social Sciences*, 18.
P. K. Feyerabend (1989a) 'Realism and the Historicity of Knowledge', *Journal of Philosophy*, 86.
P. K. Feyerabend (1989b) 'Antilogikē, in F. D'Agostino and I. C. Jarvie (eds.), *Freedom and Rationality: Essays in Honor of John Watkins*. Dordrecht: Kluwer.
P. K. Feyerabend (1991) *Three Dialogues on Knowledge*. Oxford: Blackwell, 1991 (TDK).
P. K. Feyerabend (1991a) 'Concluding Unphilosophical Conversation', in G. Munévar (ed.), *Beyond Reason: Essays on the Philosophy of Paul Feyerabend*. Dordrecht: Kluwer.
P. K. Feyerabend (1991b) 'Monster in the Mind', *The Guardian*, Friday, December 13.
P. K. Feyerabend (1991c) 'Atoms and Consciousness', *Common Knowledge*, 1.
P. K. Feyerabend (1992a) 'Ethics as a Measure of Scientific Truth', in W. R. Shea and A. Spadafora (eds.), *From the Twilight of Probability: Ethics and Politics*. Canton, Mass.: Science History Publications.
P. K. Feyerabend (1992b) 'Nature as a Work of Art', *Common Knowledge*, 2.
P. K. Feyerabend (AM3) *Against Method*, 3rd ed. London: Verso, 1993.
P. K. Feyerabend (1993a) 'The End of Epistemology?', in J. Earman et al. (eds.), *Philosophical Problems of the Internal and External Worlds: Essays on the Philosophy of Adolf Grünbaum*. Pittsburgh: University of Pittsburgh Press.
P. K. Feyerabend (1993b) 'Intellectuals and the Facts of Life', *Common Knowledge*, 2.
P. K. Feyerabend (1994a) 'Art as a Product of Nature as a Work of Art', *World Futures*, 40.
P. K. Feyerabend (1994b) 'Has the Scientific View of the World a Special Status, Compared with Other Views?', in J. Hilgevoord (ed.), *Physics and Our View of the World*. Cambridge: Cambridge University Press.
P. K. Feyerabend (1994c) 'Realism', in C. C. Gould and R. S. Cohen (eds.), *Artifacts, Representations and Social Practices*. Dordrecht: Kluwer.
P. K. Feyerabend (1994d) 'Potentially Every Culture Is All Cultures', *Common Knowledge*, 4.
P. K. Feyerabend (KT) *Killing Time: The Autobiography of Paul Feyerabend*. Chicago: University of Chicago Press, 1995.

P. K. Feyerabend (1995a) 'Philosophy, World and Underworld', and 'Science, History of the Philosophy of', in T. Honderich (ed.), *The Oxford Companion to Philosophy*. Oxford: Oxford University Press.

P. K. Feyerabend (1995b) 'Universals as Tyrants and as Mediators', in I. C. Jarvie and N. Laor (eds.), *Critical Rationalism, Metaphysics and Science. Essays for Joseph Agassi*, Vol. 1. Dordrecht: Kluwer.

P. K. Feyerabend (2000) *The Conquest of Abundance*. Chicago: University of Chicago Press.

E. Gellner (1968) 'The New Idealism—Cause and Meaning in the Social Sciences', in I. Lakatos and A. E. Musgrave (eds.), *Problems in the Philosophy of Science*. Amsterdam: North-Holland.

J-F. Lyotard (1979) *The Postmodern Condition: A Report on Knowledge*. Manchester: Manchester University Press.

P. B. Medawar (1963) 'Is the Scientific Paper a Fraud?', *The Listener*, 70. (Reprinted in his *The Strange Case of the Spotted Mice, and other classic essays on science*. Oxford: Oxford University Press, 1996).

N. Murphy (1990) 'Scientific Realism and Postmodern Philosophy', *British Journal for the Philosophy of Science*, 41.

Z. Parusnikova (1992) 'Is a Postmodern Philosophy of Science Possible?', *Studies in History and Philosophy of Science*, 23.

J. M. Preston (1994) Review of J. Dupré, *The Disorder of Things, Philosophical Books*, 35.

J. M. Preston (1995) 'Frictionless Philosophy: Paul Feyerabend and Relativism', *History of European Ideas*, 20.

J. M. Preston (1997a) *Feyerabend: Philosophy, Science and Society*. Cambridge: Polity Press.

J. M. Preston (1997b) 'Feyerabend's Retreat from Realism', *Philosophy of Science*, 64 (Proceedings).

G. A. Reisch (1998) 'Pluralism, Logical Empiricism, and the Problem of Pseudoscience', *Philosophy of Science*, 65.

J. Rouse (1987) *Knowledge and Power: Toward a Political Philosophy of Science*. Ithaca: Cornell University Press.

J. Ziman (1980) *Teaching and Learning about Science and Society*. Cambridge: Cambridge University Press.

8

Paul Hoyningen-Huene

Paul Feyerabend and Thomas Kuhn

In this chapter, I want to discuss several aspects of the relationship between Paul Feyerabend and Thomas Kuhn.[1] There are seven sections. The next section will contain biographical information about the two, especially when, where, and how these two thinkers came together. Then, in the third section, I'm going to discuss aspects of Kuhn's and Feyerabend's theories that have become trademarks. By discussing these issues one can see the greatest parallels between their theories. In the fourth section, I'm going to show the large differences that existed in the 1960s between these two thinkers, which can be seen in Feyerabend's general criticisms of Kuhn's *Structure of Scientific Revolutions*. It is important to show these differences because Feyerabend and Kuhn are often seen as sitting in the same boat, as if their opinions in the philosophy of science were identical. This, however, is not the case, because in the 1960s they could not put aside their massive ideological differences. In the fifth section, I'm going to analyze Feyerabend's direct criticism of Kuhn—specifically of Kuhn's account of normal science. In the sixth section, I'll critically discuss Feyerabend's argument against Kuhn's account. Finally, I'm going to briefly show how, in the late 1980s, Feyerabend changed his opinion about his philosophical relationship to Kuhn.

In order to expose these differences, in addition to published works by Feyerabend and Kuhn, I'm going to use two letters that Feyerabend wrote to Kuhn sometime around 1960 or 1961, in which he evaluated and criticized a draft of Kuhn's *Structure*. These two letters are quite extensive and were recently published.[2] They comprise thirty-one single-spaced pages and were found in Feyerabend's *Nachlass*. He had forgotten their existence.

1. Biographical Remarks

Paul Feyerabend, born 1924, went as a guest professor to the Philosophy Department of the University of California at Berkeley in 1958. The year after, he received a permanent position. Prior to that he worked at Bristol. Thomas Kuhn, born 1922, went as an assistant professor to the University of California at Berkeley in 1956. Later he became a full professor, primarily in the Department of the History of Science. Prior to that he was at Harvard. According to Feyerabend himself, he first read Kuhn's work in 1959. By the latest they were well acquainted by 1960, and their most intensive interaction was in 1960 and 1961. Their positions at the same university ended in 1964, when Kuhn went to Princeton. Since then, they only met a few times. Their last meeting was in June 1985 in Zurich when Kuhn, who was in Paris, was invited by Feyerabend to spend three days in Zurich. These were days of intensive academic and personal discussions. Kuhn gave a talk at the ETH Zurich to a full auditorium. The following day, there was a pleasant cruise on the Zurich lake with a dozen friends and acquaintances. Feyerabend always called this trip 'the intellectual cruise'.

Feyerabend became known noticeably earlier than Kuhn. He made a name for himself in the late 1950s, predominantly as a knowledgeable philosopher of physics who had some provocative opinions that were clearly formulated and full of sharp arguments. He traveled often, gave talks, and had many discussions. Kuhn, at this time, was only known by the inner circle of historians of science. This circle was small because in the 1950s the historiography of science had just begun to establish itself professionally in North America. In the early 1960s, Feyerabend was noticeably instrumental in helping Kuhn become more well-known among philosophers. In several talks from 1960 to 1961, Feyerabend mentioned that a book from Kuhn was forthcoming: a book that would enforce and support Feyerabend's own theoretically developed opinions by way of concrete examples drawn from the history of science. Even so, some time went by before Kuhn became generally well-known, even after the publication of his *Structure of Scientific Revolutions* in 1962. In the well-known *Encyclopedia of Philosophy*, published in eight volumes in 1967, and which has since undoubtedly been a standard reference work in the English-speaking world, there is not one hint of Thomas Kuhn. Furthermore, there is not even a single article about the history of science, although there are over 4,200 large, two-column pages. Paul Feyerabend, by contrast, was mentioned and discussed in an article about the philosophical consequences of quantum mechanics and he was the author of four articles about the physicists Boltzmann, Heisenberg, Planck, and Schrödinger.

What brought Feyerabend and Kuhn together, so that they were so often mentioned in the same breath? In those days, both had reservations about the dominant philosophical tradition in the Anglo-Saxon world—namely, logical empiricism. Both had a solid scientific background. Feyerabend had a masters in astronomy. Kuhn had a Ph.D. in theoretical physics (his supervisor was John Van Vleck, who won the Nobel Prize in 1977). Feyerabend developed his critical attitude toward the neopositivist tradition not primarily from the history of science, but from the discussion of the empirical basis of science, the so-called 'protocol sentence' debate, and from his intensive analysis of Popper. Kuhn's skepticism toward the philosophical tradition had its roots in his analysis of the history of science, which he began in 1947. It appeared to him that the *actual* history of science did not fit well with the *normative* image of science that philosophers had developed. Most important to their philosophical similarity, in the eyes of their peers at least, was the fact that they both simultaneously introduced a new concept into the philosophy of science in their extraordinarily influential works from 1962: Feyerabend in his essay 'Explanation, Reduction and Empiricism' and Kuhn in *The Structure of Scientific Revolutions.* This concept later became the center of philosophical controversy that began immediately after the publication of these works and which continues even today, without loss of strength and without prospect for consensus in the near future. The conflict about this concept is now an established part of the philosophy of science, for example, in the controversy about rationality and realism, that Kuhn and Feyerabend also sparked or substantially intensified. This controversial concept is called incommensurability.

2. Incommensurability

The concept of incommensurability turns out to be extraordinarily difficult and controversial. Feyerabend once expressed it like this: "Apparently everyone who enters the morass of this problem comes up with mud on his head". I will not discuss, in what follows, the concept of incommensurability in all its details, but only give a somewhat schematic overview—a more precise analysis would need its own essay. First, I want to show what is meant by the concept of incommensurability and why it was so provocative for the preceding tradition in the philosophy of science. Then, I will talk about the main difference between Feyerabend's concept of incommensurability and Kuhn's.

The central point of incommensurability is that theories that replace one another, separated by a scientific revolution, do not work with exactly the same concepts. On the one hand, a new theory invents new concepts and throws away certain concepts of the old theory. For example, in Newton's theory of gravitation, the concept of gravitational force is introduced and the concept of the center of the universe that was essential for the older theory was rejected as unnecessary and useless. But the more interesting cases of incommensurability are when concepts of the old theory are used in the new theory as well, but with a somewhat different meaning. For example, the concept of mass is used in Newton's mechanics and in the theory of relativity as well, but it does not mean exactly the same thing in both theories. These changes of concepts, often called 'conceptual shifts', lead to a special relationship between the prerevolutionary and postrevolutionary theory. This relationship between two theories is what the incommensurability concept is about. As I said, Feyerabend and Kuhn differ in several respects, to which I will turn in a minute, as to exactly how the concept of incommensurability is characterized. But first, I would like to mention four central characteristics of the relationship between incommensurable theories that are valid for both Feyerabend and Kuhn's conceptions.

1. Incommensurable theories are incompatible, but their incompatibility cannot be transformed into a logical contradiction. This may not knock you out of your chair, but for the received neopositivist philosophical program it was extraor-

dinarily provocative. This program contended that philosophical insight can only be accomplished by logical analysis. But according to Feyerabend and Kuhn, incommensurability cannot be completely characterized with the tools of logic—at least not with the tools of existing logic. This led them to criticize logical analysis as the exclusive tool of the philosophy of science, and to promote hermeneutic or anthropological methods for the philosophy of science as well.

2. Incommensurable theories make different claims about what exists in the world, or more sharply, what the world is. Correspondingly, one can also say, as Feyerabend and Kuhn did, that the world changes with a scientific revolution. What this means exactly is another question and to answer this question one would need at least a separate essay.

3. Incommensurable theories are not literally translatable into each other. That means that with the concepts of the one theory, one cannot formulate the other theory completely. Correspondingly, one has to learn a new language to understand the new theory—at least a group of established concepts are strangely and weirdly altered. In order to have a good command of both theories, one must, in a certain sense, become bilingual. Even if one has accomplished this, one is not able to translate one theory into the language of the other theory, because to be bilingual is not identical with having the ability to translate.

4. Comparing two incommensurable theories with respect to their capabilities is substantially more complicated than comparing commensurable theories. With commensurable theories that employ substantially the same apparatus of concepts, one compares the predictions of the two theories with respect to their empirical precision. For the single predictions of the two theories this is not problematic, because for each prediction of the one theory there is a corresponding prediction of the other. Difficulties can only arise if one theory is better than the other theory in a certain area, while the other theory is better in another area. This case does not arise if one theory is a special case of the other, and this is how theory succession in modern science was understood, at least in physics. With incommensurable theories, one cannot compare them by superimposing their separate single predictions, because of the difference in their concepts and the untranslatability of both theories or, to put it differently, because they conceptualize differently more or less the same domain. That is why certain claims of one theory cannot find counterparts in the other theory, and what is from the perspective of the one theory evaluated as a substantial achievement is from the perspective of the other theory simply not understandable.

Let me illustrate. In chemistry in the first half of the eighteenth century, combustion was understood as a procedure in which a certain substance—'phlogiston'—was set free. That is not so implausible if one observes a candle flame, or if one notices that the ash of wood is much lighter than the wood that was burned. The question arose, how heavy is phlogiston, and different methods of measuring were suggested. After the so-called chemical revolution, combustion was understood as oxidation, as a process of binding with oxygen, and phlogiston was declared not to exist. Obviously, the question about the weight of phlogiston does not make sense any longer and quantitative results of the older chemistry about it find no direct counterpart in the new chemistry. From the perspective of the new chemistry, all data about the weight of phlogiston are simply irrelevant, no matter how precise and well reproducible they were. Similarly, many claims in the new chemistry have no counterpart in the old. Although such claims must play a role in the assessment of a theory, they cannot compete directly with the corresponding claims of the other theory, and that is how the comparison of the theories becomes substantially more complex.

A certain part of the controversial discussion about the conception of incommensurability can be traced back to the fact that Feyerabend and Kuhn were often understood as wanting to deny the possibility of an achievement-oriented comparison of theories altogether. But with that, the development of science would be arbitrary, as it would lack any rational procedure of theory-choice. This interpretation does not fit with the intentions of Feyerabend and Kuhn (even if Feyerabend, in particular, often sounded this way). They primarily wanted to argue against the received and, in their opinion, oversimplified understanding of the relationship of replacing theories, where the choice of theories exclusively by comparison of their capabilities was made by the comparative evaluation of single predictions.

Now to the main differences between Feyerabend's conception of incommensurability and Kuhn's. Kuhn's conception of incommensurability has a larger scope than Feyerabend's. The reason is that for Feyerabend only *comprehensive* theories can be incommensurable and then only if they are interpreted in a *certain way*. Examples of pairs of incommensurable theories are, for Feyerabend: quantum mechanics and classical mechanics, or relativist and classical physics—all in a certain interpretation. Such comprehensive theories are part of the constitution of their objects. If one uses a theory that is incompatible with another comprehensive theory, one gets different objects and the claims of both theories can no longer be directly compared with each other. As a consequence of the restriction of Feyerabend's incommensurability thesis to comprehensive theories, there are some theory-pairs that are incommensurable for Kuhn but not for Feyerabend. For example, the Ptolemaic and Copernican theories of planets, which is for Kuhn one of the standard examples of incommensurability, is for Feyerabend, on the other hand, explicitly rejected as a candidate for incommensurability.

Now I turn to the controversy between Feyerabend and Kuhn about Kuhn's *Structure of Scientific Revolutions*.

3. Feyerabend's General Criticism of Kuhn's *Structure of Scientific Revolutions*

Feyerabend respected Kuhn a great deal. Modest as he actually was, contrary to the impression he could also give, he always thought that Kuhn was more important than himself. He began his famous 1970 essay on Kuhn entitled 'Consolations for the Specialist', contained in the Lakatos and Musgrave volume, with the following words:

In the years 1960 and 1961 when Kuhn was a member of the philosophy department at the University of California in Berkeley I had the good fortune of being
able to discuss with him various aspects of science. I have profited enormously
from these discussions and I have looked at science in a new way ever since.
(Feyerabend 1970, p. 197)

Despite this, Feyerabend had deep reservations about Kuhn's *Structure of Scientific Revolutions*. Although he acknowledged the problems Kuhn struggled with
(especially the problem of the omnipresence of anomalies in science), he often
disagreed with Kuhn's theoretical treatment of these problems. I cannot deal with
the many detailed critical remarks that Feyerabend formulated in his two letters
to Kuhn. Instead, I will concentrate on the global point of criticism with which
Feyerabend began the first letter to Kuhn and which is the central point of his
essay from 1970. Feyerabend suspects that underlying Kuhn's thinking is an ideology that "could only give comfort to the most narrow-minded and most conceited
kind of specialism". For Feyerabend, this ideology "would tend to inhibit the
advancement of knowledge" and "is bound to increase the anti-humanitarian tendencies . . . of . . . science". Such an ideology is, for Feyerabend, naturally a flagrant
target open for attack. Feyerabend, on the one hand, sees himself as a strict opponent of all dogmatic tendencies—in science, in philosophy, and later in politics
as well. On the other hand, he sees himself as a champion of humanitarian values
such as liberty and the right to develop one's individuality. Thus, he is an opponent of all forms of oppression of minorities and other cultures. With respect to
science, for Feyerabend it is "the most important question of all . . . to what extent
the happiness of individual human beings, and to what extent their freedom, has
been increased". This question is a result of his conviction "that the happiness
and the full development of an individual human being is now as ever the highest
possible value". How does it come about that Feyerabend suspects Kuhn of a
hidden agenda, an ideology with antihumanitarian tendencies?

Feyerabend's accusation has, on closer inspection, three components that I
will summarize by way of introduction before we consider them in detail. First,
Kuhn formulates, as is well-known, a certain picture of how science in the basic
disciplines typically historically develops. Kuhn's book about the structure of scientific revolutions formulates a general model of the stages of the development of
science that should be valid for basic research. There are, in other words, certain
regularities in the succession of different phases of science. Feyerabend first doubts
the historical adequacy of some aspects of Kuhn's (in this sense) schematic description of the development of science. Thus, he contradicts Kuhn in a historical
(or descriptive) respect. Second, beyond these descriptions, Kuhn discusses the
functional role that different elements of science have for scientific progress. For
example, Kuhn characterizes so-called 'normal science' as containing a certain
dogmatic tendency and explains why this dogmatic tendency is not, paradoxically,
an obstacle to the development of science, but rather conducive to it. Kuhn, thus,
assesses certain aspects of science positively by explaining their function for further
development. Secondly, Feyerabend doubts Kuhn's assessment of these elements
of science. Thus, he contradicts Kuhn in methodological (or evaluative or norma-

tive) respects. Third, in Kuhn's presentation of his theory in the *Structure of Scientific Revolutions*, the descriptive and evaluative aspects are not cleanly separated, more often descriptive and evaluative elements alternate. The reason for this from Kuhn's perspective (at least in retrospect from 1969 and in reaction especially to Feyerabend) is that under certain assumptions, the descriptive claims about the development of science indirectly imply normative claims for the successful operation of science. Feyerabend, on the other hand, sees in Kuhn's mode of presentation an insidious way of taking the reader in with his own ideology without the possibility of critical distance; thus, he criticizes Kuhn with respect to his mode of presentation. To summarize, Feyerabend criticizes Kuhn in three respects: historical-descriptively, methodological-evaluatively, and with respect to his mode of presentation.

So now to the concrete details. At the heart of these criticisms is the so-called 'normal science' that Kuhn brought into the discussion. Normal science, following Kuhn, is a phase of the development of science that is marked by the broad consensus that the scientific community shares with respect to the basic questions within their discipline. This consensus is based on concrete scientific achievements which are so convincing and which have so much heuristic potential that further research in the corresponding discipline can model itself upon them and choose problems and solutions in analogy to them. Such exemplary concrete scientific problem solutions are called 'paradigms' by Kuhn. The concrete research activity in normal science has certain similarities with a totally different activity, puzzle-solving. The given characterization of normal science has as an assumption that there is no doubt about the exemplary character of the paradigmatic problem solutions—more strongly put, that the paradigms are effective dogmatically. They are effectively dogmatic in the sense that during normal science they cannot be called into question, thus they do not have to be justified or continually tested. Instead, in the practice of normal science one assumes that they are valid. But for Kuhn, normal science leads always to scientific revolutions, where the until then valid exemplary problem solutions are replaced by new ones. This happens in phases of extraordinary science where several different theories, the old theory, improved versions of it, and totally new ones, are tried out and compared.

First, Feyerabend doubts that the existence of normal science is a historical fact. Instead, he believes that there is no temporal differentiation of, as he calls it, periods of theory proliferation and monism, but that those are ways of doing science which coexist simultaneously. Feyerabend's reasoning for this assertion is not very convincing: he just gives one example that he thinks proves the coexistence of theory proliferation and the work inside one tradition. But this example, as so often in the history of science, could also be interpreted differently. It does not seem to me to be a specific weakness of argumentation by Feyerabend, but rather an expression of the difficulty of justifying or refuting basically statistical assertions about regularities in the history of science by historical material. I will not follow this criticism of Kuhn and its inherent problems here because, by comparison with Feyerabend's following second and third points of criticism, it is marginal.

Feyerabend's second criticism concerns the evaluation of normal science. For Feyerabend, normal science is, to put it simply, a horror, just as it is for the other critical rationalists of the 1960s—especially Popper and Watkins. I group Feyerabend with the critical rationalists because in 1960–61, when he wrote the two letters to Thomas Kuhn, an essential part of his arsenal of arguments was indeed taken from the critical rationalist camp. The reason for his aversion to Kuhn's normal science is its dogmatic or quasi-dogmatic element, as Kuhn himself explicitly developed it. For Kuhn, this dogmatic element is not at all tied to a devaluation of normal science, but is functional for scientific progress. On the other hand, for a critical rationalist, a dogmatic element inside science is absolutely unacceptable. The reason is that for a critical rationalist science is, and must be, at heart critical. Science consists of producing highly falsifiable hypotheses and testing these hypotheses as rigorously as possible. One of the central tenets of critical rationalism is that if one decides not to test certain hypotheses, then one ceases to do science. If Kuhn evaluates the dogmatic element of normal science positively, he shows, in the eyes of the critical rationalist, a fundamental violation of the scientific ethos, namely to be critical and undogmatic. And for the enlightener Feyerabend, this violation is also antihumanitarian, because "[p]rogress has always been achieved by probing well-entrenched and well-founded forms of life with unpopular and unfounded values. This is how man gradually freed himself from fear and from the tyranny of unexamined systems" (ibid., pp. 209–210). That is why Feyerabend identifies Kuhn's 'normal' element of normal science as conservative and antihumanitarian. I will return to more details in the next section.

But what enraged Feyerabend even more is Kuhn's mode of presentation, as Feyerabend understood it, and this is his third main criticism. Feyerabend grants that Kuhn needs a point of view, or as he also puts it an "ideology", which provides the background for his account in the sense that it influences the interpretation of the historical facts. Without such an interpretation-generating point of view, any historical account "would be the most drab and uninteresting affair imaginable". In other words, Feyerabend allows a "methodological", that is, evaluative, point of view according to which Kuhn evaluates certain elements of science as rational and other elements as irrational. Feyerabend has especially in mind Kuhn's positive evaluation of normal science. However, he accuses Kuhn of not making this evaluative point of view explicit so that readers can take notice that there are also alternative points of view which can lead to other evaluations. Instead, with Kuhn it looks as if the evaluation of historical facts follows immediately from these facts themselves. Feyerabend summarizes this aspect of his criticism very concisely in his first letter to Kuhn: "What you are writing is not just history. It is *ideology covered up as history*." And, "It is this bewitching way of representation to which I object most, the fact that you take your readers in rather than trying to persuade them. This manner of presentation you share with Hegel and Wittgenstein." Feyerabend calls this aspect of Kuhn's account its "ambiguity": a vacillation between description and prescription. Feyerabend goes as far as to claim openly and in print in his 1970 essay that Kuhn's ambiguity is intended:

> I venture to guess that the ambiguity is *intended* and that Kuhn wants to fully exploit its propagandistic potentialities. He wants on the one side to give solid, objective, historical support to value judgements which he just as many other

people seem to regard as arbitrary and subjective. On the other side he wants to leave himself a safe second line of retreat: those who dislike the implied derivation of values from facts can always be told that no such derivation is made and that the presentation is purely descriptive.

I do not know if Feyerabend felt good about this massive and in fact slanderous accusation. In any case, in his first letter to Kuhn, he asks at the end of a similar paragraph: "Or have I perhaps completely misunderstood you?" And in the printed essay from which the above quote was taken, he pleads for support for his own interpretation because he is fortified in his opinion "by the fact that almost every reader of Kuhn's *Structure of Scientific Revolutions* interprets him as I do, and that certain tendencies in modern sociology and modern psychology are the result of exactly this kind of interpretation." This passage is surprising as Feyerabend is not known as a philosopher who is very impressed by majority opinions, not to mention using them to legitimize his own position. A bit later in the same essay, Feyerabend asks if those readers misinterpreted Kuhn—and I do not think that for Feyerabend this was a purely rhetorical question.

4. Feyerabend's Criticism of Kuhn's Evaluation of Normal Science

In any case, at the end Feyerabend leaves open the question about the mode of presentation of Kuhn's book. But their deepest point of difference is the assessment of normal science to which I now, in my fifth section, turn again. How deep this difference is can be seen by the fact that Feyerabend is obviously unable even to describe normal science in neutral, that is, nonevaluative, terms. In passages where he intends to present Kuhn's view about normal science, he describes normal science as the "most boring and most pedestrian part of the scientific enterprise" that deals with "tiny puzzles" and that is based on "the monomaniac concern with only one single point of view". Briefly put, normal science is "professional stupidity". I remark, by the way, that these characterizations invite the same criticism of Feyerabend as Feyerabend made of Kuhn—the mixing of descriptions with evaluations. Such characterizations strongly inhibit Feyerabend's ability to handle the positive function for the development of science that Kuhn attributes to normal science. But apart from his emotional reaction, he tries to disprove Kuhn's argument for the functionality of normal science for scientific development with a sophisticated counterargument. He formulates this argument by stating "three difficulties" with which Kuhn's functional argument for normal science is confronted.

Feyerabend's argument against Kuhn's positive assessment of normal science has the following structure. First, he describes Kuhn's functional argument for normal science. Then, he shows that Kuhn's functional argument is based on two presuppositions that are both untenable. Finally, he denies the historical existence of normal science. Feyerabend's argument against Kuhn's positive assessment of normal science can thus be reconstructed in four steps.

Step 1: Feyerabend's description of Kuhn's functional argument for normal science: "Normal science, [Kuhn] says, is a *necessary presupposition of revolutions*". This is a functional characterization of normal science that aims at explaining why normal science is a good thing: because it has the function of leading to revolutions. The persuasiveness of this functional characterization of normal science rests on two presuppositions, namely that "revolutions are desirable" and that "the particular way in which normal science leads to revolutions is desirable also". Steps 2 and 3 analyze these two presuppositions.

Step 2: Analysis of the first presupposition of Kuhn's functional argument, namely that revolutions are desirable. The "first difficulty of [Kuhn's] functional argument" that Feyerabend identifies consists in his contention that the desirability of revolutions cannot be founded in Kuhn's theory. According to Feyerabend, it is impossible for Kuhn to assess the changes that a revolution leads to as improvements because of the incommensurability that Kuhn himself recognizes. If a scientific revolution brings only a change but not an improvement, then there is no argument for the desirability of revolutions.

Step 3: Analysis of the second presupposition of Kuhn's functional argument, namely that the particular way in which normal science leads to revolutions is desirable. The "second difficulty of [Kuhn's] functional argument" that Feyerabend identifies consists in his contention that there is a better route to revolutions than normal science. Feyerabend thinks that according to Kuhn, scientists would do normal science "until disgust, frustration and boredom makes it quite impossible for them to go on"; they would "suddenly [give] up when the problems get too big". Feyerabend confronts this route to revolutions with an alternative. According to this alternative, revolutions should be caused by following the principle of the proliferation of theories: creating competitors to a given theory. In light of the competing theories, the difficulties of the former theory will be emphasized and simultaneously the means for repairing or getting rid of the difficulties can be seen. For Feyerabend, this procedure also leads to revolutions but, as opposed to Kuhn, in a rational way. Thus, Kuhn's normal science route to revolutions is not desirable.

Step 4: The third difficulty of Kuhn's functional argument is that the historical existence of normal science is more than doubtful. And this is the end of Feyerabend's counterargument against Kuhn's positive evaluation of normal science.

5. Critical Discussion of Feyerabend's Argument

Let me now assess Feyerabend's argument with respect to how convincing it is, which brings me to my sixth section. I will discuss the argument's four steps in turn.

Step 1: First of all and most importantly, the presentation of Kuhn's functional argument for normal science is not at all appropriate. The positive assessment of normal science by Kuhn is not at all derived from the fact that normal science leads to revolutions, as Feyerabend alleges. Rather, its leading to revolutions only makes it *acceptable* that normal science contains a certain quasi-dogmatic element. The main function of normal science is rather that it produces, in an extraordinarily efficient way, scientific knowledge, of which a certain part also survives the next revolution.

Step 2: Feyerabend's "first difficulty" is an immanent-critical argument based on a certain interpretation of Kuhn's concept of incommensurability. According to this interpretation, incommensurability implies that one can no longer talk of scientific progress through revolutions. But in truth, this implication is not part of Kuhn's theory: For Kuhn, incommensurability does not exclude scientific progress at all, but only precludes certain conceptions of scientific progress, to which the conception of the cumulative progress of knowledge as approaching truth especially belongs. The first difficulty that Feyerabend mentions is thus based on a misunderstanding of Kuhn's conception of incommensurability and is thus irrelevant. I point out, by the way, the deep irony that one of the inventors of incommensurability misunderstands the other inventor's conception with respect to a very substantial aspect.

Step 3: The "second difficulty" that Feyerabend mentions is that he sees a much better way to trigger scientific revolutions than Kuhn's normal science—namely the permanent proliferation of theories. At first, let there be no doubt that step 3, if it is correct, is not an argument that hits Kuhn directly because of the misdescription of Kuhn's functional argument. But even independently, Feyerabend's argument is problematic. The background of this argument is Feyerabend's conviction that sometimes a theory's substantial anomalies can only be discovered from the perspective of a competing theory. From the perspective of the original theory, these anomalies are invisible, according to Feyerabend. I call this thesis the *anomaly importation thesis*. This thesis has its own difficulties. They become visible when one asks in exactly what relationship the two theories in the anomaly importation thesis stand. Let us suppose that the two theories are *commensurable*. In this case, one theory can be articulated basically with the vocabulary of the other. Under these circumstances, it is difficult to see why a substantial anomaly for one theory is invisible from its own perspective, but visible from the perspective of the other theory. Let us therefore suppose things are the other way around, that the theories are *incommensurable*. Then one has to explain how it is possible that the defenders of the second theory who see a substantial anomaly for the first theory can ever convince the defenders of the first theory of this circumstance. Because of the incommensurability of the two theories and the resulting ontological disparities between them, the proponents of the first theory can always claim that difficulties of their own theory that are diagnosed from the perspective of the second theory are irrelevant. I do not claim that these difficulties with Feyerabend's argument are insurmountable, but in any case, they present a challenge for his position.

Step 4: The "third difficulty" of the functional argument concerns the question of whether normal science really exists historically. But that is not truly a *difficulty* of the functional argument (however it is formulated). Rather, the historical reality of normal science is a *presupposition* of the functional argument. It only makes sense to ask the question of whether normal science is functional for scientific progress if normal science really exists. The functional argument would not be rendered problematic by the nonexistence of normal science, but it would become superfluous. The "third difficulty" according to Feyerabend, is thus no *difficulty* for Kuhn's functional argument, but at most a difficulty with this entire theory. But, as I mentioned earlier, Feyerabend's historical argument against the existence of normal science is not very convincing.

Let me briefly add a word about Kuhn's own defence of the existence of normal science against Feyerabend and the other Popperians. Kuhn attempted to show that the existence of normal science necessarily follows from the existence of scientific revolutions. This argument assumes that scientific revolutions exist, and this assumption Kuhn shares with his Popperian critics and can thus be used unproblematically in the given context. Revolutions presuppose normal science, according to Kuhn, in the sense that there must be stages *between* revolutions, thus nonrevolutionary stages of science—if talk about scientific revolutions or upheavals is to make sense.

However one assesses the strength of Feyerabend's arguments against normal science, Kuhn felt misunderstood by Feyerabend on the decisive points, and he found Feyerabend's arguments against normal science entirely insufficient. First, Feyerabend himself reports this candidly at the beginning of his 1970 essay. According to Feyerabend, Kuhn had often said this to him in their discussion in the 1960s: "On all these points my discussions with Kuhn remained inconclusive. More than once he interrupted a lengthy sermon of mine, pointing out that I had misunderstood him, or that our views are closer than I had made them appear" (ibid., p. 198). Second, in many places of Kuhn's published reaction to Feyerabend's essay from 1970 Kuhn's disappointment with the arguments against his theory is obvious. There Kuhn goes so far as to say that the reaction of his critics leads him to postulate the existence of two Thomas Kuhns. One Kuhn had published a book with the title *The Structure of Scientific Revolutions* with certain theses. The second Kuhn wrote a totally different book, but with the same title. And this book contains, according to Kuhn, many theses that are totally incompatible with those from the first book. It is from this second book that his critics, Feyerabend included, often cite.

6. Feyerabend's Reconciliation to Kuhn in the Late 1980s

In the 1960s, when Feyerabend wrote his essay, he did not allow his opinion that he understood Kuhn properly to be changed. Also, in 1977, he reinforced his conviction that he contested the idea of normal science from the beginning. But another ten years later, his tone changed. After he read a habilitation thesis about Kuhn, he wrote in 1988 in the revised English edition of *Against Method*: "[H]aving been made aware of the great complexity of Kuhn's thought I am not at all sure that our differences are as great as I often thought they were." In an essay that came out one year later, Feyerabend even went farther. He wrote there that the "ideas [in the present essay] are very similar to, and almost identical with,

Kuhn's ... later philosophy". He reinforced this opinion in the third English edition of *Against Method* of 1993. In a posthumously published 1994 review, Feyerabend discussed a reconstruction of Kuhn's theory as a "surprisingly coherent and powerful system of thought". Thus, it seems that Feyerabend reconciled with Kuhn in the last years of life. In this ironic way, the somewhat superficial but widespread opinion that these two extremely influential thinkers of the second half of the twentieth century were really close in their thoughts is justified, after all.

Notes

1. I wish to thank Eric Oberheim for translating this essay from the German.

2. 'Two Letters of Paul Feyerabend to Thomas S. Kuhn on a draft of *The Structure of Scientific Revolutions*', *Studies in History and Philosophy of Science*, vol. 26, 1995, pp. 353–387.

Reference

P. K. Feyerabend (1970) 'Consolations for the Specialist', In I. Lakatos & A. Musgrave (eds.), *Criticism and the Growth of Knowledge*, (Cambridge: Cambridge University Press), pp. 197–230.

9

Elisabeth A. Lloyd

Feyerabend, Mill, and Pluralism

1.

Paul Feyerabend had a reputation, among many, for being antiscientific, irrationalist, antimethodological, antireason, a relativist about evidence, and an epistemological anarchist. More generally, he frequently anchors the 'extreme relativist' end of many a comparison in philosophy of science. The slogan 'Anything Goes', though, has certainly captured the philosophical imagination; as I have emphasized before (Lloyd 1996), this slogan is frequently misinterpreted as being Feyerabend's methodological recommendation for conducting scientific research. Not so.

As Feyerabend himself said: "*'anything goes' does not express any conviction of mine, it is jocular summary of the predicament of the rationalist*":

> if you want universal standards, I say, if you cannot live without principles that hold independently of situation, shape of world, exigencies of research, temperamental peculiarities, then I can give you such a principle. It will be empty, useless, and pretty ridiculous—but it will be a 'principle'. It will be the 'principle' 'anything goes'. (1978, p. 188; his emphasis)

Thus, Feyerabend's slogan was essentially a *reductio* against a certain form of rationalism, rather than a statement of his own positive view. I now suspect that one reason underlying the standard misreading of the slogan involves one of Feyerabend's genuine, positive, beliefs—specifically, his defense of the value of a proliferation of views and methods, and his insistence on the tolerance that must accompany such proliferation.

Feyerabend consistently traces his arguments for the central importance of proliferation of views and methods—and the appropriate attitudes of openness and tolerance—to John Stuart Mill's essay, *On Liberty*. When introducing Mill's ideas in his most extensive discussion of them, Feyerabend emphasizes and endorses

Mill's view that, as Feyerabend puts it, "pluralism is supposed to lead to *truth*" (Feyerabend 1981b, p. 67, his emphasis). Or, as Feyerabend put it in *Science in a Free Society*, "the only way of arriving at a useful judgement of what is supposed to be the truth, or the correct procedure, is to become acquainted with the widest possible range of alternatives. . . . The reasons were explained by Mill in his immortal essay *On Liberty*. It is not possible to improve upon his arguments" (1978, p. 86).

2.

There is, with Feyerabend, a pervasive difficulty of interpretation: his writings frequently exhibit a dialectical structure, some of them explicitly appearing in the form of dialogues. It is thus sometimes difficult to identify many of the views that Feyerabend articulates as being those that he would, in some other context, defend. Those who read his slogan, "Anything Goes", as "Paul Feyerabend's Positive Methodological Program" ran into just this difficulty. There is thus a serious question whether Feyerabend's regular appeals to Mill are best read as straightforward endorsements, or as moves adopted in particular battles, in which Feyerabend takes on the assumptions and standards of his opponents in order to beat them on their own turf (one of his favorite strategies).

I believe that Feyerabend's appeals to Mill's views were, in fact, genuine endorsements: Feyerabend never distanced himself from Mill's position, and he used it repeatedly in arguments aimed against other views. For instance, among his rather obsessive criticisms of Karl Popper, Feyerabend emphasized that the good parts of Popper's philosophy of science, which include his selectionist process, and the emphasis on proliferation and criticism, were all actually Mill's views (Feyerabend 1981a, pp. 141–142). I do not wish to pursue Feyerabend's attempts to undermine Popper's originality—preferring, as I do, the position that the good ideas in Popper were articulated and better defended by Charles Peirce and by John Dewey—but I do think that the general questions about the pluralism of methods and views are highly relevant to today's philosophy of science.

Specifically, we philosophers of science are working today within a context that includes the widespread philosophical rejection of any foundational doctrine of pure empirical content or sense-data; the acceptance of some constrained but very real embeddedness of evidence in theory; and the repeated failure of all attempted formulations of a set of methodological rules.

In addition, many thinkers have independently come to endorse some variant of a generally evolutionary picture of scientific inquiry, in which cultivation of variation—of interests, theories, and methods—is complemented by a variety of selection mechanisms, to produce a process of scientific change. Feyerabend's endorsement of Mill's claim amounts to a claim that the process of scientific change, when it occurs within the preferred context of a plurality of views and methods, is one that leads toward truth; it does not simply wander around the space of possibilities, as some have described the biological evolutionary process as doing, nor does it pass from the views of one powerful ruling class to another.

Moving away from this claim in its most general form, when we consider Feyerabend's own arguments regarding specific candidate views and methods, we seem to run into trouble. Feyerabend did, after all, defend witchcraft, astrology, faith-healing, Chinese medicine, and other nonscientific ways of understanding the world. It might seem that Feyerabend defeated his own purposes by doing so. Arguing that the widest variety of views and methods should be tolerated, pursued, and even nurtured, by those seeking knowledge of the world, Feyerabend could be seen—and has been seen—as displaying disregard, or even contempt, for empirical evidence, for sound reasoning, and for established scientific results, by his enthusiastic defenses of marginal, mythical, or magical systems, explanations, and practices.

To read him this way is to misunderstand what he is doing. To interpret Feyerabend's arguments in the light of John Stuart Mill's essay *On Liberty*, in contrast, is to illuminate the stage that Feyerabend took himself to be on, and to understand properly the role he so frequently elected to play.

3.

Let us turn to John Stuart Mill. We concentrate on Mill's long essay, *On Liberty*, co-written with his wife, Harriet Taylor, and first published, as a small book, in 1859. The relevant section is chapter 2, "Of the Liberty of Thought and Discussion", in which Mill denies that people, "either by themselves or by their government" should "attempt to control the expression of opinion" (Mill [1859] 1977, p. 229). "The peculiar evil of silencing the expression of an opinion", writes Mill, in a passage frequently quoted by Feyerabend,

> is that it is *robbing the human race*. . . . If the opinion is right, they are deprived of the opportunity of exchanging error for truth: if wrong, they lose, what is almost as great a benefit, the clearer perception and livelier impression of truth, produced by its collision with error. (Mill [1859] 1977, p. 229; my emphasis)

Mill argues for freedom of expression of opinion on four distinct grounds:

> First, if any opinion is compelled to silence, that opinion may, for aught we can certainly know, be true. To deny this is to assume our own infallibility. Secondly, though the silenced opinion be an error, it may, and very commonly does, contain a portion of truth; and since the general or prevailing opinion on any subject is rarely or never the whole truth, it is only by the collision of adverse opinions that the remainder of the truth has any chance of being supplied. Thirdly, even if the received opinion be not only true, but the whole truth; unless it is suffered to be, and actually is, vigorously and earnestly contested, it will, by most of those who receive it, be held in the manner of a prejudice, with little comprehension or feeling of its rational grounds . . . fourthly, the meaning of the doctrine itself will be in danger of being lost, or enfeebled, and deprived of its vital effect on the character and conduct: the dogma becoming a mere formal profession, inefficacious for good, but cumbering the ground, and preventing the growth of any real and heartfelt conviction, from reason or personal experience. (Mill [1859] 1977, p. 258)

Regarding the first reason, Mill argues:

> the opinion which it is attempted to suppress by authority may possibly be true.
> Those who desire to suppress it, of course deny its truth; but they are not infallible
> ... To refuse a hearing to an opinion, because they are sure that it is false, is to
> assume that their certainty is the same thing as absolute certainty. All silencing
> of discussion is an assumption of infallibility. (p. 229)

For Mill, such an assumption of infallibility is not merely a moral problem—it
has the effect of reducing the opportunity for humanity-at-large to ascertain truth.

The second argument, in which the kernel of truth in minority opinions must
be preserved—for the good of everyone—is particularly interesting. The situation
is one where "conflicting doctrines, instead of being one true and the other false,
share the truth between them" (p. 252). Given, then, the "partial character of
prevailing opinions",

> every opinion which embodies somewhat of the portion of truth which the com-
> mon opinion omits, ought to be considered precious, with whatever amount of
> error and confusion that truth may be blended. No sober judge of human affairs
> will feel bound to be indignant because those who force on our notice truths
> which we should otherwise have overlooked, overlook some of those which we
> see. Rather, he will think that so long as popular truth is one-sided, it is more
> desirable than otherwise that unpopular truth should have one-sided asserters too;
> such being usually the most energetic, and the most likely to compel reluctant
> attention to the fragment of wisdom which they proclaim as if it were the whole.
> (Mill [1859] 1977, p. 253)

Mill's third and fourth arguments are more subtle: he insists that, even if the
received opinion is true, those holding it need—for their own good, and the good
of everyone—to be "vigorously and earnestly contested". With no discussion challenging the view, people may believe the truth, but they will not believe it in a
reasonable manner, that is, it will simply be a prejudice. Worse yet, Mill argues,
without the presence of conflicting opinions, people cannot really understand the
meanings of their own beliefs. These arguments rely on Mill's vision of the nature
and operation of human judgment.

The fact that human errors are corrigible is, Mill says, a quality of the human
mind which is "the source of everything respectable in man either as an intellectual or as a moral being" (p. 231). "He is capable", Mill writes,

> of rectifying his mistakes, by discussion and experience. Not by experience alone.
> There must be discussion, to show how experience is to be interpreted. Wrong
> opinions and practices gradually yield to fact and argument: but facts and argu-
> ments, to produce any effect on the mind, must be brought before it. Very few
> facts are able to tell their own story, without comments to bring out their mean-
> ing. The whole strength and value ... of human judgement ... [depends] on the
> one property, that it can be set right when it is wrong, [and] reliance can be
> placed on it only when the means of setting it right are kept constantly at hand.
> In the case of any person whose judgment is really deserving of confidence, how
> has it become so? Because he has kept his mind open to criticism of his opinions
> and conduct. Because it has been his practice to listen to all that could be said

against him; to profit by as much of it as was just, and expound to himself, and upon occasion to others, the fallacy of what was fallacious. Because he has felt, that the only way in which a human being can make some approach to knowing the whole of a subject, is by hearing what can be said about it by persons of every variety of opinion, and studying all modes in which it can be looked at by every character of mind. No wise man ever acquired his wisdom in any mode but this; nor is it in the nature of human intellect to become wise in any other manner. The steady habit of correcting and completing his own opinion by collating it with those of others, so far from causing doubt and hesitation in carrying it into practice, is the only stable foundation for a just reliance on it. (p. 232)

This view of informed judgment forms the basis for Mill's conclusion that, however true an opinion may be, "if it is not fully, frequently, and fearlessly discussed, it will be held as a dead dogma, not a living truth" (p. 243).

Furthermore, the only way to bring the reasons and arguments supporting a true belief into genuine contact with the believer's mind, is to hear the arguments of adversaries "in their most plausible and persuasive form; he must feel the whole force of the difficulty which the true view of the subject has to encounter and dispose of; else he will never really possess himself of the portion of truth which meets and removes that difficulty" (p. 245). In fact, Mill writes, "if opponents of all important truths do not exist, it is indispensable to imagine them, and supply them with the strongest arguments which the most skilful devil's advocate can conjure up" (p. 245).

4.

This is where Paul Feyerabend comes in. I want to suggest that seeing Feyerabend as attempting to enact and embody these views of Mill, provides a valuable interpretive framework for his more peculiar actions and extreme views. In particular, I am suggesting that Feyerabend was compelled by Mill's account of the importance of rational discussion, and insisted—often without the consent of his interlocutors—on engaging in the sort of discussion which, for Mill, served as the foundation for rational opinion and conduct.

I do not mean to imply that Feyerabend kept his intentions secret—actually, after reading Mill carefully, it seems that Feyerabend left stage directions all over his contributions, asides announcing to his audience: 'I am now playing a believer in a minority view in order to elevate the intellectual and moral level of this discussion'. Mill wrote, "I confess I should like to see the teachers of mankind endeavouring to provide a substitute for [a diversity of opinions]; some contrivance for making the difficulties of the question as present to the learner's consciousness, as if they were pressed upon him by a dissentient champion, eager for his conversion" (Mill [1859] 1977, p. 251). I am suggesting that Feyerabend often assigned himself the role of such a champion, and in doing so, was attempting to enact Mill's vision of the most productive form of interchange.

For instance, regarding his impassioned defenses of astrology, Feyerabend writes: "My use of examples from astrology should not be misunderstood. Astrology bores me to tears. However it was attacked by scientists, Nobel Prize winners among them, without arguments, simply by a show of authority and in this respect deserved a defence" (1991, p. 165).

In addition, Feyerabend carefully explained the motivation behind the form of many of his arguments:

> [A]ttempts to retain well-entrenched conceptions are criticized by pointing out that the excellence of a view can be asserted only after alternatives have been given a chance, that the process of knowledge acquisition and knowledge improvement must be kept in motion and that even the most familiar practices and the most evident forms of thought are not strong enough to deflect it from its path. (1981a, p. xi)

Moreover, like Mill, Feyerabend saw the beneficiaries of his defenses of unpopular views as the observers of the interactions. Mill wrote, regarding the importance of the advocacy of minority opinions: "It is not on the impassioned partisan, it is on the calmer and more disinterested bystander, that this collision works its salutary effect" (Mill [1859] 1977, p. 257). Compare this to what might appear to be one of Feyerabend's more perverse—but typical—perspectives:

A: Are you an anarchist?

B: I don't know—I haven't considered the matter.

A: But you have written a book on anarchism!

B: And?

A: Don't you want to be taken seriously?

B: What has that got to do with it?

A: I do not understand you.

B: When a good play is performed the audience takes the action and the speeches of the actors very seriously; they identify now with the one, now with the other character and they do so even though they know that the actor playing the puritan is a rake in his private life and the bomb-throwing anarchist a frightened mouse.

A: But they take the writer seriously!

B: No, they don't! When the play gets hold of them they feel constrained to consider problems they never thought about no matter what additional information they may obtain when the play is over. And this additional information is not really relevant . . .

A: But assume the writer produced a clever hoax . . .

B: What do you mean—hoax? He wrote a play, didn't he? The play had some effect, didn't it? *It made people think*, didn't it? (Feyerabend 1991, pp. 50–51, emphasis added)

In sum, what at first (and second) glance may appear to be a scattershot and unprincipled approach to issues of scientific knowledge by Feyerabend is—plau-

sibly—actually quite principled and unified, once we take seriously what Feyerabend says about Mill.

5.

It could be objected that Feyerabend has misapplied Mill; after all, Mill's essay is
usually remembered today for its defense of freedom of speech—especially the
expression of heterodox religious views—against governmental authority. Mill's
work on how scientific reasoning ought to proceed is presented in his *System of
Logic*.

I find it difficult to deny that Mill, himself, saw the sciences, especially what
he called the physical sciences, as operating within a smaller, perhaps more protected, set of expectations and norms; he wrote about these norms in his *System
of Logic*, the first edition of which was published in 1848, and which he continued
to revise and expand throughout his life. There is a sense in which Mill treated
scientific topics as somewhat special cases of discussion, controversy, and resolution, for all of the obvious and sensible reasons.

In contrast to Mill's discussions in his *Logic*, which concern reasoning about
facts and evidence, many of Mill's points in *On Liberty* are illustrated through
explicitly religious cases, in which there was no promise of factual evidence to
which appeal could be made. Hence, one could argue that Mill, in *On Liberty*,
is responding to extremist and sectarian religious views, cases in which there is no
evidential means for settling the issues at stake. Should we follow Feyerabend in
applying Mill's arguments favoring pluralism—which seem to be addressing primarily views that remained outside the purview of scientific inquiry—to scientific
knowledge? Does Feyerabend's appeal to Mill amount to a misapplication of
Mill's argument?

I do not think so. Contrary to its reputation and use today, Mill stated explicitly that *On Liberty* is not primarily about the limits of government interference
(p. 305). He was much more concerned about social norms, claiming that protection is needed against "the tyranny of the prevailing opinion and feeling" and that
society sometimes "practices a social tyranny more formidable than many kinds of
political oppression, since, though not usually upheld by such extreme penalties,
it leaves fewer means of escape, penetrating much more deeply into the details of
life, and enslaving the soul itself" (p. 220).

The great bulk of the essay concerns how individual people (not governments)
ought to respond when confronted with opinions and forms of life that are strange
or disagreeable to them. The central point regards the inestimable value—to individuals and to society as a whole—of the existence and nurturance of a wide
variety of ways of life. Given what he sees as individuals' natural, unreflective
intolerance toward those who see and do things differently than they do (p. 227),
Mill emphasizes the importance of reducing social and cultural sanctions against
those who espouse minority views of any kind.

Perhaps even more crucial is the cultivation of attitudes and skills that, as
Mill sees it, are necessary for the genuine flowering of human intelligence and

creativity. These attitudes include tolerance, but Mill also demands much more; he advocates "social support for nonconformity" (p. 275). In fact, Mill argues, "In this age, the mere example of nonconformity, the mere refusal to bend the knee to custom, is itself a service. Precisely because the tyranny of opinion is such as to make eccentricity a reproach, it is desirable, in order to break through that tyranny, that people should be eccentric . . . the amount of eccentricity in a society has generally been proportional to the amount of genius, mental vigour, and moral courage which it contained" (p. 269). Ultimately, Mill claims that "diversity of character and culture" is what has led to human progress (p. 274); "[t]he only unfailing and permanent source of improvement is liberty, since by it there are as many possible independent centres of improvement as there are individuals" (p. 270).

Most importantly for Feyerabend's adaptation of it, this value of the diversity of opinion was not, for Mill, restricted to areas such as politics and religion, for which no factual appeals were available; Mill explicitly included the sciences in his recommendation of "absolute freedom of opinion and sentiment on all subjects, practical or speculative, scientific, moral, or theological" (p. 225). Moreover, science is included not solely on the basis of a principle of liberty, but on the grounds that pluralism is needed for attaining truth:

> [O]n every subject on which difference of opinion is possible, the truth depends on a balance to be struck between two sets of conflicting reasons. Even in natural philosophy, there is always some other explanation possible of the same facts; some geocentric theory instead of heliocentric, some phlogiston instead of oxygen. (p. 244)

I conclude that there is, therefore, no misapplication of Mill's views in Feyerabend's applying them to the scientific context.

6.

Thus Mill's primary preoccupation in *On Liberty* was with positive means of cultivating human flourishing, and not with delineating the limits of legitimate government interference. There is something radical, though, in Feyerabend's use of Mill—a use that is consonant with Mill's own views, but which is in tension with most twentieth-century philosophy of science—namely, the aim of recontextualizing the sciences (and philosophical discussions of the sciences) back into their larger roles in human ways of living. Mill was including modern scientific approaches in the range of ways of living that individuals in a free society may choose among. Because people differ, different ways of life will be judged valuable by them:

> If a person possess any tolerable amount of common sense and experience, his own mode of laying out his existence is the best, not because it is the best in itself, but because it is his own mode . . . Such are the differences among human beings in their sources of pleasure, their susceptibilities of pain, and the operation on them of different physical and moral agencies, that unless there is a corre-

sponding diversity in their modes of life, they neither obtain their fair share of happiness, nor grow up to the mental, moral, and aesthetic stature of which their nature is capable. (Mill [1859] 1977, p. 270)

Furthermore, Mill defended the notion of each individual pursuing their own path on the basis of its value to the rest of human society. Under conditions of freedom, Mill writes,

human beings become a noble and beautiful object of contemplation; and as the works partake the character of those who do them, by the same process human life also becomes rich, diversified, and animating, furnishing more abundant aliment to high thoughts and elevating feelings, and strengthening the tie which binds every individual to the race, by making the race infinitely better worth belonging to ... There is a greater fulness of life about his own existence, and when there is more life in the units there is more in the mass which is composed of them. (p. 266)

What has sometimes been seen as Feyerabend's "antiscientific" attitude is, I suggest, more appropriately interpreted as his taking this aspect of Mill's thought very seriously. The central point is not the moral one, that people should not be made to live in a way that they did not freely choose; it is that individuals truly can be the best judges—better even than any scientific experts—of which way of living is better for them, and that we could all benefit from these judgments. Consider the following passage:

B: ... Ever since people were discovered who did not belong to the circle of Western culture and civilization it was assumed, almost as a moral duty, that they had to be told the truth—which means, the leading ideology of their conquerors. First this was Christianity, then came the treasures of science and technology. Now the people whose lives were disrupted in this manner had already found a way of not merely surviving, but of giving meaning to their existence. And this way, by and large, was much more beneficial than the technological wonders which were imposed upon them and created so much suffering. 'Development' in the Western sense may have done some good here and there, for example in the restriction of infectious diseases—but the blind assumption that Western ideas and technology are intrinsically good and can therefore be imposed without any consultation of local conditions was a disaster. (Feyerabend 1991, p. 74)

Feyerabend applies this point equally to more specific areas such as medicine, in which scientific expertise should not—for scientific purposes as well as moral ones—be taken as a sufficient reason to reject other forms of expertise relevant to these aspects of life. Feyerabend writes, regarding traditional Chinese medicine, for example:

[A]lternative medical systems are often parts of entire traditions, they are connected with religious beliefs and give meaning to the lives of those who belong to the tradition. A free society is a society in which all traditions should be given equal rights *no matter what other traditions think about them.* A respect for the opinions of others, choice of the lesser evil, chance of making progress—all these things argue in favor of letting all medical systems come out into the open and freely compete with science. (1991, p. 75; his emphasis)

Here, Feyerabend is not appealing to a simple libertarian right of people to live as they wish; he also provides suggestions regarding promising areas of scientific research, focusing on areas in which Western scientific medicine has weaknesses that may be corrected by the introduction of Chinese techniques and understandings.

7.

In summary, I have suggested following Feyerabend's own advice and interpreting his work in light of the principles laid out by John Stuart Mill. A review of Mill's essay, *On Liberty*, emphasizes the importance Mill placed on open and critical discussion for the vitality and progress of various aspects of human life, including the pursuit of scientific knowledge. Many of Feyerabend's more unusual stances, I suggest, are best interpreted as attempts to play certain roles—especially the role of "eccentric defender of unpopular minority opinion"—that are necessary to fulfilling Mill's conditions for rational exchange and optimal human development.

I wish that I could ask Feyerabend just how explicit and extensive his intentions were to enact Mill's context for rational judgement. I missed my chance, but this is merely my painful loss. I have attempted here to characterize our loss—as philosophers of science—in terms that I believe cast a rather different light on Feyerabend's obstreperous and, to many, slightly insane defenses of unpopular viewpoints. He can be seen as trying to provoke us into protecting us from ourselves and to highlight and actually enact the most fundamental principles by which human life is actually improved. For this, I would like to thank him.

Note

I am particularly grateful to Mathias Frisch, Ben Hansen, Maria Merritt, Ina Roy, and Eric Schwitzgebel, for valuable discussion and suggestions in preparing this chapter.

References

P. K. Feyerabend (1975) *Against Method: Outline of an Anarchistic* Theory of Knowledge.* London: Verso.

P. K. Feyerabend (1978) *Science in a Free Society.* London: Verso.

P. K. Feyerabend (1981a) *Realism, Rationalism & Scientific Method: Philosophical Papers,* Vol. 1. Cambridge: Cambridge University Press.

P. K. Feyerabend (1981b) *Problems of Empiricism: Philosophical Papers,* Vol. 2. Cambridge: Cambridge University Press.

P. K. Feyerabend (1991) *Three Dialogues on Knowledge.* Oxford: Basil Blackwell.

E. A. Lloyd (1996), 'The Anachronistic Anarchist', *Philosophical Studies,* 81, pp. 247–261.

J. S. Mill ([1848] 1973–74) *A System of Logic,* ed. J. M. Robson, Toronto: University of Toronto Press.

J. S. Mill ([1859] 1977) *Essays on Politics and Society,* in J. M. Robson (ed.), *Collected Works of John Stuart Mill,* Vol. 18. Toronto: University of Toronto Press.

10

J. N. Hattiangadi

Two Concepts of Political Tolerance

Introduction

The intellectual authority of science is now at a much lower ebb than at any time since the Second World War. By the "intellectual authority of science" I mean how its value is regarded by the intelligent reading public, and among the other esoteric academic professions. "Intellectual authority" does not mean that everyone takes its conclusions to be true, of course, though among the less knowledgeable it may have this effect. In the Middle Ages, for instance, philosophy enjoyed great intellectual authority. In more recent times, this authority has gradually slipped. Mathematics certainly has seen its authority enhanced since the Middle Ages. The intellectual authority of natural science has not been overtaken by another subject, as that of philosophy was taken over by its precocious offspring, physics, three hundred years ago. All the same, in recent decades, science has seen its intellectual authority slip, which is the background for this essay. The knowledge that it has slipped is in the evidence of the ease with which critiques of science are being published.[1]

One part of the story of the decline in the intellectual authority of science has to do with the writings of Paul Feyerabend and Thomas Kuhn, which have contributed by their influence over the last four decades to this state of affairs.[2] The upshot of the writings of both Feyerabend and of Kuhn, in different ways perhaps, is that their work provides us with a better understanding of the context of scientific activity than the positivist theories that they criticized. The decline of logical positivism has left a vacuum where a clear picture of science was once available for all to attack. The study of intellectual contexts, which could have provided the alternative, has not been taken up by subsequent students of science in quite the numbers that it has deserved. Consequently, relativists have had a rather easy time of it, ridiculing positivism, and scoring points for their own views, which can take account of contexts.

In recent years, there have been two somewhat different movements that have grown quite large, each of which undermines the intellectual authority of science: one is called the "social constructivist" critique (and the "postcolonial" critique can be subsumed under it), and the other, the "postmodern" critique. The third critique, the feminist critique, sometimes joins hands with one of them, but it need not do so, and so it does not necessarily undermine the intellectual authority of science.

The background of these two anti-intellectual critiques has been the rediscovery of the close connection between epistemology and political theory. Although this connection has been a recurring theme in philosophy from its very beginning, it has become more visible as a connecting thread more recently due to the controversial and influential work of K. R. Popper in *The Open Society and Its Enemies*, and of M. Foucault in *The Order of Things*.[3] Feyerabend's *Science in a Free Society* is another work in the same vein, which sees epistemology in the context of political theory and vice versa.

In the first part of this chapter, I will offer a brief account of the relativism characteristic of the new debunking of science, as it developed from an old reaction to the Enlightenment ideal of an objective science and its social consequences. Some regrettable outcomes of this philosophical reaction have been evident as it spread from Germany, where it was first formulated as a doctrine, to the rest of the world, in the two alternate guises of nationalism and (international) socialism. Part two will extend this analysis back in time by identifying a fundamental problem with classical liberal theory, as it was developed by John Locke. This problem exacerbates the controversy over objectivity, and it is an epistemological source of the divergence between the liberal and democratic conceptions on one side and communalist views on the other of the nature of tolerance. I will illustrate this with an influential and troubling case of judicial appeal in India.

In this section, I will also outline Karl Popper's approach to this problem and explain why Feyerabend's critique of Popper's theory of knowledge was a telling one, both for the theory of science and for Popper's defense of liberal democracy. Finally, in the last section, I will propose what I take to be a simple way of regaining liberal democracy from Feyerabend's critique, which seems to me to meet the challenge of all the contemporary critiques.

When there is a disputed issue, and the two sides to the dispute take part in a discussion of the issue, neither side can dictate to the other that their own partisan presuppositions must be accepted as given. In agreeing to participate in a discussion of a certain kind, an appeal to truth and reality are implicit, as are certain methodological constraints that may seem congenial to all the parties in the dispute. The presuppositions of the intellectual disagreement are in that sense not presupposed by (or "relative to") any of the contrasting opinions on the issue under dispute (though one does not wish to deny that they may well presuppose other assumptions being made by all of the contestants at the time).

This simple thought can be used to develop a whole typography of intellectual situations that throw up rules for contestants without any legislation or interference from above. Our task here is to address the more general issues that have been raised in recent years, following the recent eclipse of the intellectual authority of science. This account is enough to rescue objective truth and reality from the relativist who wishes to decry them.

1.

After Napoleon first won and then lost control of Europe on the battlefield, the opinion of those loyal to the French Revolution and its principles was tinged by a feeling of regret. They regretted that he had failed to liberate Europe from its age-old unjust and unjustifiable customs, as they saw them. Although they might have felt embarrassed that an emperor had emerged from the populist effort to defend liberty, equality and fraternity, there was, no doubt, chagrin that with the Battle of Waterloo, the liberation promised by the Encyclopedists had been delayed, or even thwarted. However much they wished that Napoleon had not gained power or that, having gained it, he had won his battles, they could no longer hope that the French Revolution and the American War of Independence could be made available to the larger part of Europe, or to the rest of the world.

The liberation of Europe had become necessary according to the revolutionaries because they thought that they had discovered the secret of the hold that the haves had over the have-nots: the ruling aristocracy controlled the knowledge that allowed them to benefit at the expense of the have-nots. Diderot's *Encyclopédie*, with its "reasoned dictionary of the arts and sciences" was an effort, beginning with the first volume in 1751, to make knowledge widely available.[4] Its effect was far more dramatic than the Chambers *Cyclopedia* of 1729, which it tried to emulate in French. The landed aristocracy of Europe was expected to fight the dissemination of practical knowledge on which their well-being depended, and so it seemed that it became necessary that the rebellious intellectuals fight back. Whether or not it was necessary to fight back, that is how the rebels saw it. Perhaps the landed aristocracy saw it thus, too, because in 1749 Diderot was imprisoned for some of his writing, and in 1759 the *Encyclopédie* was banned. D'Alembert's preface to the Encyclopedia extolled the great and unprecedented success of Newtonian (or perhaps Copernican) science in overthrowing ancient dogma and superstition.[5] D'Alembert himself resigned from the editing after its ban. It seemed to the rebels that this great new ("modern") movement could not be extended to the reform of all the other irrational customs and traditions until after the European aristocracy had been defeated in military battle. The rebels felt keenly that they had been driven to battle by the intransigence and power of vested interests. The march of knowledge (as in the motto "Let Truth Prevail") had its critics, of course: but those who disagreed were cast as reactionaries who were either manipulators, or deluded, or paid lackeys of manipulators.

Among the landed aristocracy, however, a different view prevailed. To them the French Revolution seemed to be an outbreak of mob rule, much as Plato had depicted Athenian democracy in the last days of Socrates, only worse for being more vulgar and egalitarian than anything in ancient times. Napoleon was regarded as a bounder, who had risen with luck to high military rank, but without the pedigree, and so found that he could not handle power when it fell to him. It seemed to them reasonable to assert that only those born to and bred for power have the necessary discretion in the use of it to be able to keep it and to use it effectively. Of course there is, in this, an implicit critique of egalitarianism, which inevitably throws up commoners to lead commoners (the blind to lead the blind). But the bulk of the "reactionary" analysis about the Napoleonic era (during its existence and after) is devoted to the megalomania of the upstart Napoleon. This constitutes much of what British historians had to say about this historical period, for instance.[6]

It is primarily in Germany that the response to the Philosophic Emperor was itself a philosophy. In a sentence, the new philosophy was that Napoleon's so-called "war of liberation" was really only French imperialism, or the attempt to subjugate the German way of life to French will. If in France it seems that a certain German custom is unjust, it does not follow that the custom is unjust. The principles used to judge whether justice is done in any given case differ from nation to nation. To apply French theories of justice to German feudal relations is not only unjustifiable, but it is moreover a much greater injustice than anything that the Germans might do to one another. True liberation could not come to Germany by the occupation of the fatherland by Napoleon's troops, with their alien culture, on the pretext of its appeal to universal reason. Rather it would come by the unification of all German-speaking peoples through an acquiescence to one German law, which would be just law because it would be their own.

The appeal to abstract principles has since then always been suspected of being a ruse to impose one's will upon another, or, in short, "imperialism".

Heinrich Heine noted that one of the roots of this philosophic analysis is to be found in the philosophy of Kant, even though Kant was one of the greatest defenders of the need to liberate all of humankind. But as an enlightenment philosopher, Kant had rationally worked out that being moral is a matter of doing one's duty because it is one's duty. Thus if I obey God's commandments, I must do so because it is my duty to do so, and not merely because it is God who commands me. Thus, says Heine, God is no longer the ruler of the universe: "Hear ye not the bells resounding? Kneel down. They are bringing the sacraments to a dying God."[7]

Before the critique of German idealists (which is, in all fairness, pregnant with the "social constructivist" and the "postmodern" critiques of more recent times) made its mark, the emergence of the individual rational agent as the final court of appeal in all questions of morality was an inspiring idea. Individual rationality went even beyond God's law (though it may turn out to be one's duty to obey God, after all), and it may be summarized in the doctrine, "the rational man is autonomous". Of course, this presupposes that there is a reason that is universal in the sense that any individual of any nation or culture will partake of it.

On the other hand, what if Herder is right? Herder argued that no thought or argument exists, except in some language or other, and that a language is characteristic of a community. "Our mother-tongue embodies the first universe we see, the first sensations we feel, the first activities and pleasures we enjoy. Secondary ideas of time and place, of love and hate, and all the flaming impetuous thought of youth are perpetuated by it. This perpetuation of thoughts and feelings through language is the essence of tradition".[8] Thoughts, then, are characteristic of those communities in which they arise. Is reason contextually limited? If we adopt the view that thought (reason) makes for autonomy, but that thoughts are (reason is) characteristically communal, it follows that it is communities that are, truly speaking, morally autonomous. This view, that it is communities that are morally autonomous, I shall call "communalism" with its main subcomponent, which I call "nationalism".[9] Nationalism implies that no nation may interfere in the internal affairs of another, for to do that is to try to impose one community's reason upon that of another, which is imperialism. Imperialism violates the right of a community to be treated as an end in itself, and not merely as a means, just as Kant had once asserted of individuals. Reason, as Kant envisaged it, is by definition imperialist, of course, because reason is defined as universal to humans, and those who think that there is no such thing take this doctrine of universality to be a mere cover for the imposition of one's will upon another's.

The main argument for maintaining that what is reasonable in one context is not so in another is the view to be discussed below, namely, "epistemological relativism". Different people have different customs. The urbane individual is not shocked at the most cruel customs of others, for they are cruel only when taken out of context. Every judgment must be made in context, including judgments concerning the injustice of an action, custom, or institution. If it can be shown on rational grounds that even the most universal tenet of reason is not indifferent to context, then the philosophy of nationalism cannot be argumentatively bested.

Surprisingly, it was not in Germany that the first nationalist unification took place, but in Italy. Germany followed, and then unification spread to South Africa, Turkey, and shortly to the rest of the world. If the instant sympathy that plain people in the world have felt for this critical analysis of Napoleon's adventures, which was developed to its highest form in German philosophy, is a sign that there is something right about nationalism, then nationalism has much truth to it.

It is an unhappy truth, if true. For though nationalism enjoins communal autonomy, (as Kedourie notes) it does not identify which communities are to be or become autonomous.[10] In fact, communalism, the more general doctrine, is even more vague about what constitutes a community. Herder himself recognized the difficulty of this problem as it applied to the dominant role he assigns to culture as a determinant of language and reason. "Nothing," Herder wrote, "is more indeterminate than this word, and nothing more deceptive than its application to all nations and periods".[11]

Hegel's pioneering researches into communities and their interplay is also ambivalent over this very issue. Sometimes he favors professional or work communities, which are better known these days as "classes". On other occasions he prefers land-based communities, which may be called nations.[12] Two great political movements, national communalism, better known as "national socialism", and international communalism, known just as "socialism", or "international socialism", owe some allegiance to Hegel for their philosophic roots.

National as well as international socialism before Marx had difficulty with the identity of communities. Every "liberated nation" of the last century has had difficulty with groups within it who have claimed communal autonomy on racial, linguistic, religious, or geographic grounds, and demanded liberation from the imperialism of the larger, dominant group in the nation. The Boer war was the first civil war to presage the future: what happens when the struggle for unification divides communities (leading to civil war) when the issue is not easily decided, unlike the unification of Italy and Germany, where, initially at least, the issues seemed to be clearer? Let us not forget the fate of Armenians during the triumph of Kemal Ataturk and the young Turks. Nor the saga of the unification of Ireland, which perhaps is at last on the verge of being pacified. Horrendous cruelty and insensitivity, unknown in the annals of "noble" warfare of days past, has been witnessed. The search for what constitutes a community in Eastern Europe, or "balkanization", as it is known, led to the assassination of Archduke Ferdinand, and the outbreak of the First World War. It spread to Spain, where the dominance of art over science made that country an easy prey to the communal question.[13] The Second World War, or the ensuing chapter of the continuing civil war in Europe, soon followed. Now every part of the world is engulfed in the search for what constitutes a truly autonomous moral community.[14] The idea that some things are plainly unjust has lost all political favor. In some regions of the world the philosophical search for the definition of community is more vicious than others, but no one is spared this search for the moral and epistemological autonomous unit.

International socialists thought that their scientific approach avoided this deadly affliction. Karl Marx tried to give a well-defined theory of classes to counteract the nebulousness of socialist theory on the identity of morally autonomous communities. The theory was very attractive to many. In practice, however, it has not been borne out. If it seemed clear in the Bolshevik revolution what constitutes the proletariat, it soon became less clear. The peasant proletariat of Mao's revolution, and the Solidarity communal movement in Poland, and the vast differences that exist between workers and their representatives in the histories of socialist states, in their different ways suggest that there is no clear answer to the identity of community (class) in international socialism either. It seems, rather, that the issue is deeply and ideologically troubled rather than a tactical matter in need of practical political action (as most Marxists try to interpret it).[15] Communal strife is raising its head not only in Russia and the former socialist states, but in liberal democracies as well, so powerful are the intuitions governing the doctrine that I have called "communalism". Even those who defend liberal democracy to their last breath will concede rights to communal groups in practice. This is a great confusion, and a dangerous practice. But is there a better alternative?

Given how important theories of knowledge and of truth have been in this account of the recent history of civil warfare, it is important to extend the account to an examination of the epistemological questions, and the questions concerning truth, that have prompted these events. Feyerabend's book *Science in a Free Society* cannot be read freed from these concerns.

To conclude this section I need only repeat Napoleon's famous maxim that the pen is mightier than the sword. Napoleon himself thought, no doubt, that his battles in Europe were the aftermath of the penmanship of Voltaire and Condorcet, of D'Alembert and Diderot. Perhaps he saw his subjugation of his neighbors by the sword as merely a prelude to the true liberation of Europe by the pen. It seems that even Napoleon did not see that the pen is mightiest when fighting the sword, especially when the sword appears to have severed the hands that write. For the reaction of the German pen to Napoleon's sword has led to a movement of communal liberation so powerful that no part of the world has been left untouched.[16] In fact, in most parts of the world, both the French Revolution and its abstract ideals are remembered (especially in its bicentennial year)[17] largely as a precursor to celebrate a communal (national) liberation, which is the last ironic comment on the impetuous and unfortunate Napoleon Bonaparte.

2.

It might seem that the role of epistemological relativism is rather nebulously stated in the historical survey preceding this section. But it is not difficult to find the source of the relation between relativism and the communalist attack on classical liberal doctrine.

John Locke, in each of his epistles concerning toleration, takes the difference between what is known and what is not as the foundation of his political philosophy. Civil and criminal law, based on reason, is part of what one knows, but religious matters are in the realm of faith, and therefore are not known to us. Locke suggests that in matters that are beyond our knowledge (as in religious affairs), one must be tolerant. Thus, he asserts that "the magistrate's power extends not to the establishing of any articles of faith, or forms of worship, but by force of his laws. For laws are of no force at all without penalties, and penalties in this case are absolutely impertinent; because they are not proper to convince the mind."[18] Moreover, "if truth makes not her way into the understanding by her own light, she will be but the weaker for any borrowed force violence can add to her".[19] A person may pursue any religious sect as a matter of faith in a place of worship that is appropriate, provided that it does not violate the laws that are part of what we know on the basis of reason. In fact, such is the influence of Locke's idea that the word *faith* has come to have "religion" as one of its central meanings.

The difference between what we do know and what we do not is the very foundation of British liberal thought since Locke. Locke's *Essay Concerning Human Understanding*, which is a study of the scope and limits of human knowledge, is an integral part of his political philosophy.[20] Unless we can say what it is that we do know, and what the limits of our knowledge are (i.e., what sorts of thing we cannot know), we cannot adopt a satisfactory (Lockean) principle of toleration. Locke's attack on innate ideas and his strict dependence on empiricism effectively banish religion from the realm of knowledge into the realm of faith. It seemed to Locke that the best examples of knowledge (e.g. mechanics) that were then available fit nicely into his scheme for empirical knowledge. In the eighteenth century Locke's empiricist foundations for liberal political thought and Newton's extraordinarily successful mechanics and optics were regarded as part of the same basic fabric of British opinion. Few who said they were Newtonian went on to dissent from Locke's empiricist foundations for natural philosophy, David Hume being one of those very few. Unfortunately for Locke's political philosophy, even though Newtonian science seemed an excellent example of knowledge, and the faith of every Protestant sect a counterexample to knowledge, it proved impossible to show that any system of civil and criminal law is derivable from rational considerations alone. But as long as Newtonian science continued to be regarded as a piece of established empirical knowledge, it seemed reasonable to think that with effort one could work out the true system of justice.

Although Locke was attempting to open up a space for tolerance, the principle that anything established by the use of reason may not be open to disagreement can be interpreted as an intolerant doctrine. For if I satisfy myself that anything contrary to my wishes is contrary to reason, and I happen to be an absolute monarch, (perhaps even for good reason as I have determined by argument), then I can be as intolerant as I wish on the grounds of the principle of toleration. As an atheist, I may, for instance, outlaw all religion as offensive to reason. Locke, after all, forbade Hobbists (atheists) as well as Papists (Roman Catholics) from political participation of any kind. Clearly there is a problem here with this kind of selective toleration, though Locke may have good reasons for making these particular exceptions.[21]

To illustrate how this philosophical problem has played itself out on the issue of communal relativism, I shall select a case from India; not a case of communal strife, of which there are many, for the First World War has by now reached every corner of the globe;[22] but a case involving judicial and parliamentary decisions, conducted in comparative peace and with due deliberation and regard for reason, where philosophic qualm, if it exists, might have had its chance to surface. Indeed philosophical discussions have followed, and so have many political consequences, leading up to the election of the party in power today.

The Shaha Bhanu case was appealed to the Supreme Court of India a few years ago, in which a Muslim woman appealed the imposition of a pittance as her share of the inheritance of an estate, as determined by the Shariat law of Islam. According to this law, a widow receives one-eighth of a husband's estate upon his death, and no more, the balance being divided among other living relatives. From the estate of a deceased parent a daughter will receive one-half of the share a son receives. Something seemed wrong in this scheme of inheritance to the complainant and will perhaps seem so to some others as well. Ignoring details of the case, I report only that the appeal to the Supreme Court to overrule the application of Shariat Law was based on principles of natural equity or justice, as well as on constitutional grounds. The Supreme Court of India found for the woman complainant.

Muslim religious leaders vehemently protested this travesty of God's law at human hands. The prime minister of India hastily addressed Parliament and legislated that the legitimacy of the Shariat law must be observed in such cases (where Muslims alone are concerned), and thus overruled a landmark decision of the Supreme Court of India.

Commentators accused the prime minister of pandering to votes (Muslim votes, one must not forget, are very important to any secular party in India). Defenders of the prime minister accuse the critics (who are mostly lawyers, jurists, journalists, educators, and of course, all opposing politicians) of being anti-Muslim, and therefore of being communalists. The matter is, of course, not so simple. Now that Muslims are allowed this degree of autonomy in mistreating women of their community, Hindu fanatics have demanded parallel concessions: they have become increasingly dissatisfied with secular laws which, for example, prohibit women from voluntarily and heroically jumping into their husband's funeral pyre (a Hindu rite of dubious origin, which was accorded the status of a Hindu tradition when the British passed a law, still on the books since the nineteenth century, forbidding it). Is it not perfectly just that if Muslims can mistreat women in their community, as judged, of course, by false Western standards, then Hindus can do likewise, as judged by the same false Western standards? In fact all the religions can take turns appealing for more justice for their quaint but morally sacrosanct customs. Women who appeal to Western ideals of morality should read Rorty or Feyerabend. Which would they rather have—false Western justice, or their own comfortable communal practices?

I give this case not because Indians are more vicious than other countries to their women. On the whole, they are. The issue at hand here, however, is not that, but it is a question whether Hindus have the same right to mistreat women as Muslims, under the principle that no community should interfere in the rights of another. Never mind that there is a great difference between laws of inheritance and of the right to life. The issue is a matter of philosophic principle. There is a political issue that remains unresolved: what constitutes true toleration? Respecting all people, and their customs, for what they are, or demanding that all people be subject to reason?

There are evidently two contrasting concepts of tolerance in politics today, and they must not be confused with one another. Is toleration of Muslim custom not commendable? Are the critics of the new legislation not communalist troublemakers? On the other hand, if toleration is to exclude actions that run contrary to what a reasonable person would regard as inequitable, then it is not the Supreme Court that was intolerant. Then it is the government that has capitulated to the communal forces which ravage the country now in the grip of civil discord. We need to decide which of these two is the right point of view. To decide which is the case, one must choose between two principles of toleration—the liberal and the relativist. The liberal principle says that in any basic conflict of opinion the use of our reason will usually tell us the correct answer, failing which we must be tolerant of dissent. It leads to the intolerance of the dogmatist, at times, as we have noted. The relativist principle tells us that in case of basic conflict of opinion, there is no resolution, so we should plan to tolerate or overcome the opponent. (It is not argument but *realpolitik* that determines the outcome of disagreement). Given epistemological relativism, the principle of liberal democracy since Locke is indefensible, and vice versa.

There are relativists who take the stand that, speaking from within their view, relatively, they are liberal democrats (Rorty, for instance). This does not invalidate my remark that epistemological relativism undermines liberal democracy. For the relativist liberal democrat may claim that though there is no intercommunal rational justification for the received legal doctrines of a given community, such doctrines are nevertheless regarded as justified in that community and are distinguished from all matters which may remain open to question for members to debate in the same community. But this only adopts the words "liberal democracy" while emasculating them of all meaning. To see this, we can see that Hitler, Papa Doc Duvalier, or Ferdinand Marcos could each have claimed to be a liberal democrat in some sense of that phrase. Each supported a system of law that may be construed at some time as being justified in the community, by the methods of justification communally acclaimed by that person; each allowed dissent outside of those issues that were communally settled in the political way that was appropriate at the time.

The relativist liberal democrat may rejoin by pointing to the role of elections, freedom of association, a free press, right to political office as all part of what is intended by "relativist liberal democracy". But without a nonrelativist epistemology, this is just eyewash, for Marcos (or any other dictator) can rejoin that different communities produce the same effect in different ways, and that his Philippines was just as much a liberal democracy as, let us say, the United States or Canada, except that the communal forms that the institutions took were slightly different. There was complete freedom of the press, for instance, so long as the press acted within the laws and ordinances, written and understood, of the Philippine state. American laws are not more democratic, just different. Are the United States and Canada different? Do they let their prisoners free out of a spirit of tolerance for crime?[23]

If we are to distinguish liberal democracies from other forms of government at all, then we have to recognize that if epistemological relativism is true, then a liberal democratic critique of any society is insupportable as far as we can know it. For any liberal democracy to be possible, there must be standards apart from those communally allowed. But where can these reside? How can they transcend the community in which they are enshrined, if at all?[24]

The last question raised needs to be reinterpreted by absolutist and relativist in their different ways before it is admissible. (As stated, the relativist might insist that the use of the word *true* begs the question, for instance). If we assume that the issue has been appropriately reinterpreted by each, we can address it only after noticing that in recent years, the growth of science seems to demonstrate the failure of the classical liberal point of view.

We have seen that the liberal point of view requires a clear distinction between what is known and what is a matter of faith. Justice is decided on grounds that are known, while one tolerates faith because it falls outside of the scope of knowledge.

In this century, under the influence of the overthrow of Newtonian theory, not once but twice, and under the influence of acute analyses undermining the empiricist claims of the justification of Newtonian knowledge, we have largely abandoned the idea that in science we are given an abstract view of the world that may be considered true and established. Instead of the truth of Newtonian theory we speak of the success of science. Instead of the demarcation of knowledge from faith, we speak of the demarcation of science from metaphysics. In all this the main feature of our new quest is to come to terms with the fact that scientific theories are ephemeral, whether they be interpreted realistically as by Popper, or instrumentally as by Duhem. Given this surprising feature of science, new interest has arisen in the history of science, as distinct from mere biography of scientists. For it seems that history fashions science, after all.

If we look at what the history of science has taught us in these eighty odd years, the main conclusion to be drawn for our purposes would seem to be that science is not a timeless category with timeless characteristics, but is itself a growing intellectual tradition of the last three hundred years. Science has its own traditions and its own institutions. There is no timeless method for establishing Newtonian theory, which makes it free from reasonable doubt, once and for all. If there are methods at all, they evolve, as does our abstract understanding of the world. (This was Feyerabend's favorite hobbyhorse).

The view that nothing can be known except as part of some historical tradition or other will be called 'historism'.[25] Historism seems to imply the falsity of liberal theory. It cannot be denied that historism has been associated with epistemological relativism. As we have seen, it seems to lead to the collapse of liberal theory, if there is no timeless standard of truth. My own contribution in this chapter will be to outline a way of accepting historism without accepting relativism, postmodernism, communalism (or any doctrine which threatens liberal thought under another name).[26] I shall suggest such a way, and then suggest this as an interpretation of science that accords with all the critiques of Feyerabend (and those of several others) and saves toleration as well. In order to do this, let us review why Popper's defense of liberal democracy (and of truth and of knowledge) are severely undermined by Feyerabend's critique of them.

There is a defense of liberal democracy that seems not to assume that there is anything that we do know for sure in the sense that Locke required. This is the analysis of science of Sir Karl Popper, and his defense of liberal democracy in his works, *The Poverty of Historicism* and *The Open Society and Its Enemies*.[27] His analysis of science emphasizes the tentative nature of all theory. In fact it proposes to make its empirical falsifiability, rather than verification, verifiability, confirmation, or confirmability a defining characteristic of science. He interprets science as the unending quest for truth, and sees the institutions of science as designed to maximize testing and criticism to allow science to grow. Scientific statements, Popper claims, "always retain the character of tentative hypotheses, even though their tentativeness may cease to be obvious after they have passed a great number of severe tests."[28] And,

> The result of tests is the *selection* of hypotheses which have stood up to tests, or the *elimination* of those hypotheses which have not stood up to them, and which are therefore rejected. It is important to realize the consequences of this view. They are these: all tests can be interpreted as attempts to weed out false theories—to find the weak points of a theory in order to reject it if it is falsified by the test. This view is sometimes considered paradoxical; our aim, it is said, is to establish theories, not to eliminate false ones. But just because it is our aim to establish theories as well as we can, we must try to falsify them. Only if we cannot falsify them in spite of our best efforts can we say that they have stood up to

severe tests. This is why the discovery of instances which do not confirm a theory means very little if we have not tried, and failed, to discover refutations.[29]

This prompts Popper to suggest a theory of liberal democracy as the attempt to avoid repeated error, especially disastrous error, by maximizing criticism, no matter what other traditions or customs we may have. He can, on this basis, defend the freedom of the press, of assembly, of free elections, and most of the institutions of a modern democracy, including, surprisingly enough, the two-party system.

> But the only way to apply something like scientific method to politics is to proceed on the assumption that there can be no political move which has no drawbacks, no undesirable consequences. To look out for these mistakes, to find them, to bring them into the open, to analyze them, and to learn from them, this is what a scientific politician as well as a political scientist must do. Scientific method in politics means that the great art of convincing ourselves that we have not made any mistakes, of ignoring them, of hiding them, and of blaming others for them, is replaced by the greater art of accepting the responsibility for them, of trying to learn from them, and of applying this knowledge so that we may avoid them in the future.[30]

Popper never explicitly addresses the question in all its generality of how we can have any enlightened judgment of the injustice of many human arrangements, assuming historism in all its generality. He appeals instead to his own faith in the egalitarian enlightenment principles of the eighteenth century. It is not enough to know what Popper (or we) might like to believe, but to find a response to a critic of Locke's who says: "If we have democratic traditions, there is no justice or injustice other than what a community democratically decides. If, as often happens, a democracy tramples on some of those least able to defend themselves, then this is not unfortunate and unavoidable but quite just."

To argue for faith where Locke requires knowledge, in the matter concerning justice, makes each one of us the final arbiters in deciding what is just. If, in the end, it is faith we must rely upon to defend the enlightenment conception of justice, we have given in to the relativist point of view just where it counts most. For different people have different faiths, for historical reasons. But if we become historists, are we then not relativists? How can we then continue even to have any faith in liberal democracy once we know that it is only faith and not reason on which it is based?[31]

Feyerabend's critique of Popper's theory of science, and by extension the critique of his theory of liberal democracy, was very telling. Popper's suggestion is that in science theories are criticized (in particular, we attempt to falsify them). But these recommendations are, of course, methodological rules. Popper suggests that liberal democratic theory can allow a free discussion of all ideas if only we accept a particular methodology. But methodologies can also be disputed. We may find, and Feyerabend thought he found, that the methodology appropriate for Galileo was not Popper's but Galileo's own; for Descartes, Descartes' own, and for Newton, Newton's own. Methodology, like the abstractions of the French Revolution, is historically situated. (Scientists are "opportunists", as Feyerabend argued). If he is right, then Popper's claim for a methodology that applies to all scientists at all times is just hopelessly inadequate. It is equally hopeless, it would seem to follow, that we should find one true democratic method for all societies at all times.

There is a particular reason why Feyerabend thought that methodology appropriate to one person in one situation may not be appropriate to another in another: he thought he could argue effectively to show that scientific theories do not come as discrete statements that are examined one by one, but as whole systems, or, as he called them "comprehensive structures of thought". His preoccupation, in later life, was to show that a person in such a comprehensive structure of thought not only produced a set of facts particular to the structure, but rules of procedure (or methodological standards) as well.[32] If he is right, then Popper's defense of liberal democracy cannot succeed, for his methodology and liberal democracy itself must be historically situated. The appeal to a single standard of rationality cannot be successfully defended, he thought.

> [I]nteresting research in the sciences (and for that matter in any field) often leads to unpredictable revisions of standards though this may not be the intention. Basing our judgement on accepted standards the only thing we can say about such researches is therefore: anything goes.[33]

Feyerabend thus reintroduces his most famous slogan, about which he complained of being misunderstood.[34]

"Note," he continues,

> the context of the statement. 'Anything goes' is not the one and only 'principle' of a new methodology, recommended by me. It is the only way in which those firmly committed to universal standards and wishing to understand history in their terms can describe my account of traditions and research practices. . . . If this is correct, then the only thing that the rationalist can say about science (or about any interesting activity) is: anything goes.[35]

Feyerabend goes to great lengths to show that he does endorse standards here and there, but not everywhere. Does this lead to relativism? In another part of the same work he says, "Relativism is often attacked not because one has found a fault but because one is afraid of it,"[36] and a little later:

> That the appeal to truth and rationality is rhetorical and without objective content becomes clear from the inarticulateness of its defense. In Section I we have seen that the question 'What is so great about science?' is hardly ever asked and has no satisfactory answer. The same is true of other basic concepts. Philosophers inquire into the nature of truth or the nature of knowledge, but they hardly ever ask why truth should be pursued (the question only arises at the boundary line of traditions—for example it arose at the boundary line of science and Christianity). The very same notions of Truth, Rationality, Reality that are supposed to eliminate relativism are surrounded by a vast area of ignorance.[37]

His opinion on science and its place in a free society is captured in the following section heading: "Science is one Ideology among many and should be separated from the State just as Religion is now separated from the State", and he continues by stipulating that "a free society is a society in which all traditions have equal rights and equal access to the centers of power".[38]

In making traditions so central to his account of a free society, he loses sight of the individual. In the Shaha Bhanu case, or on the issue of the practice of widow burning, he would want the traditions to be well represented. Although Feyerabend's love of democratic and liberal institutions seems to be strong, as evidenced in this book, he has here run into the central problem dealt with in this chapter, and by making traditions equal has given the relativist conception of tolerance a strong argument.

But the trouble is that traditions come in many different forms, and there are good and bad traditions. The decision whether a tradition is good or bad, while it cannot be settled independently of traditions (nothing can) is not itself part of a tradition. Indeed what constitutes equal access to power is itself differently understood in different traditions, and hence makes nonsense of his account.[39]

Feyerabend talks of traditions, where Hegel talked more generally of history, but the issue remains the same: can we be historists without being relativists? If the answer is yes, then the new conception of science without foundations is compatible with liberal democracy; if the answer is no, it is not. We have to find a niche where intellectual standards can reside outside of communities if liberal democracies are intellectually defensible.

3.

There is no such thing as "relativism", "irrealism", or "antirealism". There are only realists, of this or that kind. The nomenclature commonly used by various critics of certain forms of realism is misleading, because nobody actually denies the reality of everything whatsoever. The relativist who denies realism is only denying the reality of some things and asserting the reality of others. For instance, the contextualist might make all existence relative to context, but would not want to consider that contexts are relative to themselves. Relative only to themselves, contexts become absolute, which implies that they do not exist according to the relativist. If there were no contexts, then all other existence would become absolute. The claim of internal realism, that nothing is real but what is called 'real' in a language, cannot be applied to the language itself. For a language relative only to itself is absolute, and if relative to another leads to the same question again. It might be possible to construct an endless imaginary chain of languages inside one another to avoid this difficulty, but it would still lack all plausibility as a description of epistemological processes. Making everything relative to traditions makes the existence of traditions and their distinctness from one another an article of nontraditional faith. To make so much of tradition is, in the end, to fall outside every tradition we know.[40]

If we thus take the relativist as a species of realist, we find that the difference between ordinary realism and the form that is advanced by the relativist lies in this: the relativist claims that all that exist are minds, or languages, language games, or human mental communities, traditions or some such interpretative substrate. All else that one might ordinarily take to exist is cast as a projection of one of the interpretative reals. But it is not at all difficult to show that any such view is flawed if we pay attention to the theory of what is real that is allowed or covertly assumed therein. A linguist will have a poor theory of our language, and an idealist a poor theory of the human mind. Having made it the only entity, and an interpretative one at that, it becomes incoherent about this one reality. Wittgenstein noted that philosophy was like tracing the framework of a picture, where the framework itself never comes into view because it is against the framework that we see anything at all. This is quite apt, for Wittgenstein's linguistic idealism makes language itself a complete mystery, whose outlines he traced obsessively over and over again. What is its origin? How does it evolve and improve? Why does it have these and not those biological features? No questions like this can even be addressed until one abandons linguistic idealism and adopts a more common form of realism.

Of course, each interpretative view can explain everything else just as well as the ordinary realist, because that is how relativism is designed, namely, to reinterpret what the realist comes up with. If the realist comes up with something new with great effort, a slight change in interpretation suffices to mirror whatever is new into the interpretative scheme, with ease. It has, as Russell said of postulation, all the advantage of theft over honest toil. In a sentence, the relativist is able to mirror everything within the interpretative scheme only because that scheme itself is emasculated of all reality and modified beyond all recognition.

It is therefore not my task to attack relativism, either the idealist version or the linguistic version, for its inconsistency or its inability to do other things. I shall take it here that it is inadequate as an empirical description of whatever it is that it describes as real relative to which everything else is reinterpreted. If at all, my attempt is to chart out the limits of relativism, that is to say, why we need to be moderate in our relativism and allow some allowance for truth, reality, and reason, however modestly.

The postmodern critique of science, truth, reality, and reason as oppressive, which follows Feyerabend, has a point: scientists can be authoritative and domineering, like intolerant priests of a religion, in some instances; truth can be in the service of the powerful; reality may be how the established make sure that the others pay obeisance to them; reason may be a way to regiment the underprivileged. But there is also another use of the concept of truth: where a little voice can hope to persuade many, it can be a source of inspiration against the powerful; science can, and has often in the past, been skeptical of those in power, and has undermined irrational authority; reason can be a subversive tool as well as a tool for domination. Postmodernists often borrow Feyerabend's flourish in attacking science as overbearing, but fail to note that in fact he also said that if the debunking of science were to become fashionable, then he would have to change his strategy. Postmodernism can be no less oppressive than truth and reality and the rest. The claims on behalf of supposed oppression also can be oppressive.

The issue we face, however, is not one of political tactics, which are in any case subservient to political goals, which may themselves be questioned. My concern is to show how a historist about science can nevertheless avoid relativism. The key to my suggestion is to undermine two ideas that are assumed without further question by most philosophers most of the time: (1) that our ideas, statements, judgments, thoughts, or propositions come in "lumps" or as "wholes"; (2)

that all truth is significant. Each of these concede too much to the relativist, which may even seem to give in to relativism. But, surprisingly, making these two concessions together allows us to accept historicity without becoming relativists.

(1) The holism that is common among philosophers in modern times began, perhaps, with Kant: in his attempt to show how abstract mathematical knowledge of the world is possible, even though induction could never yield such knowledge from perceptions, he constructed a scheme whereby the most fundamental features of the world are presupposed in the very act of perception. Thus, our physics is possible because its highest results are just the presupposition of every human being whatsoever. It is, moreover, the consistency of the worldview that we have which allows us to discover the mathematical underpinnings of our perception of things in the world. Kant's idea presupposes two things at once, it seems: first, that we have a coherent worldview; and that it is the underpinnings or presuppositions of this worldview that give the general form of dynamics. Secondly, his view is that this is universal to all humans–a natural consensus, it would seem.

Modern relativists follow Kant in supposing that there are presuppositions in all of our thought. But they do not think that these are universal, but rather universal to a group of similarly minded individuals. They do believe that these presuppositions are coherent worldviews, though they concede that there may be disagreements within a group on matter of detail. It is assumed without further analysis that the presuppositions that are assumed in common are the most important, the most fundamental, and the most fiercely defended parts of the coherent system of the world that constitutes the worldview.

The hypothesis that what is taken to be true in a community is a coherent whole can have no examples in its favor, since we are always in disagreement over any fundamental issue that we can actually name. Following Kant, however, a distinction is drawn between the mundane level of fact and the more profound level at which presuppositions operate, and it is only the latter that forms a coherent whole and that defines a community relative to which truth is defined. Since mundane disagreement is always going to be deceiving, a hidden convergence of opinion will emerge. There is never any way of isolating what these statements are, which are coherent systems of thought that we are all presupposing at a given moment.

There is little doubt that we presuppose things in all that we do, all the time. But the idea that there is a set of thoughts we presuppose that forms a coherent system is a miracle. Given how hard it is to make a trifling little subsystem consistent with all the facts as we know them, it seems an incredible assumption that, with little effort, we can come upon one coherent set of presuppositions about the world after another, just like that. Everything that we know about our presuppositions in all the situations in which we examine ourselves suggests that this not so. It is true that we presuppose a great deal. But all the evidence points to the fact that each one of us brings a very personal set of presuppositions on many issues, and that in any case many of these are not clearly consistent with one another. The assumption that we make coherent presuppositions depends on the fact that we have systems of classification (categories, as Kant envisaged them) that must presuppose a coherent worldview to give us a comprehensive and unambiguous classification of all things. But we do not possess any such classification! We may have some such classifications that work tolerably well most of the time, but a little reflection quickly shows us that we have to revise our classifications. The reason why intellectual effort is valued at all among all human societies is precisely because our systems of presuppositions, wherever they can be laid bare, show signs of incoherence, which we must correct from time to time to understand the world a little better.

But the really troubling assumption of the relativist is the assumption that what is presupposed and is coherent is also significant. This seems to me to be just trivially false.

(2) The thesis that all truth is significant is very much part of the Enlightenment view, according to which the first task of the scientist is to collect all the facts that can be collected. This seems like a reasonable task given Bacon's neo-scholastic conception of the universe (of a limited number of forms, and form engendering forms). But once it became clear that the most abstract features of the universe were still open to interpretation, it became impossible to suggest that there is a theory-independent way of specifying facts. Once we recognize that our inquiry is historist in nature, we see that the facts that are significant to one will appear pointless to another community or in another epoch. At first sight this seems to lead directly to relativism.

Let us grant that the relevant facts are determined by the nature of the opinions held within a community. If the relativist were to be challenged, we would have to show that at least some opinions are based on the facts. But this will seem not to help us, for the facts are preselected by the community for their relevance. But it is not truth that is relative to community, but relevance or significance. "Significant truth" is consequently relative. If we identify truths as worthy of study when they are significant, we may well come to the conclusion that all the truths we examine are relative. But we must note that it is at least not true in principle that truth is relative because significance is relative.

If we turn to the first point, we noticed that no science has a history without its controversies. We saw there that it is not accurate to think of scientific communities as proceeding from one consensus to another, in accordance with what has been called "a punctuated equilibrium" model. If anything, the better model is dialectical, though history is too chaotic for any simple dialectical story to be even prima facie accurate.

Now the main suggestion that I wish to make is that importance in a community of scholars depends not on what is agreed upon, but on what is controversial. If relativism is the true picture, then it would have to be reinterpreted as "relativism-to-a-controversy". The idea of relativism-to-a-controversy immediately makes relativism innocuous. In a controversy, the controverting parties oblige each other to look at unpleasant or unsavory facts. The very fact that two who disagree will discuss the issue presupposes a truth not in the possession of either party. What is important for an atomist includes the difficulties faced by the holist, and what is important for the holist includes difficulties faced by the atomist. If we take many such points of view in typical competition for scientific attention, the result is a relatively nonpartisan emergence of significance and a healthy respect for facts.

In my outline of an argument to escape relativism while accepting the historism of science, I may seem to have forgotten the big issue, of liberal democracy and its apparent failure. But in fact I have not. It often happens in a carefully argued controversy that the arguments seem technical, and proposed for their own sake, when in fact their true significance lies in their implication for some controversial matters. My arguments are, of course, no different.

Feyerabend's challenge that there are no universal standards, but that they are all situated does not imply relativism concerning truth, because standards may be situated historically, but they are not relative to any one tradition, point of view, comprehensive structure of thought, or the like. They are all born of disagreement and agreements to disagree. Partisan standards by their nature cannot decide an issue, though they may be useful for the exercise of political power. In civilized societies, the pluralism of perspectives and the ability to disagree and to continue to disagree over a great many years and generations has given rise to a fine tradition for rationality, truth, a conception of reason and, in recent years, of science. These have been friends of the poor and the downtrodden just as often as they have been the instruments of the rich and powerful. They have contributed greatly in the recent three centuries to egalitarian and democratic movements of the world, but of course this does not have to be their effect in all situations at all times. (To be fair, the relativist movement has had a salutary effect on breaking up colonial and imperial domination of peoples, too).

I have reduced the difference between relativism and the liberal democratic perspective to two principles of toleration: (1) that one must be tolerant of the views of other groups because they have epistemic and therefore moral autonomy (Feyerabend's principle), and (2) that one must be tolerant of those with whom we disagree in our own group because the truth is not established (Popper's principle).

I have argued that the inability to identify what is a natural group (community, nation, tradition) is not only a political tragedy of modern times, but a reason to abandon relativism on epistemic grounds. Finally I argued that if we examine the history of science, we will find that disagreements within groups develop into science, or into other fine intellectual traditions. If I am right, then it is possible for groups to be tolerant of internal disagreement. It is this possibility that the relativist must deny, concerning the "important" matters presupposed. We note that in any intellectual disagreement in which parties engage in a discussion (which makes it internal, as we define the term), a more than relative truth is invoked by the participants, as a mutually regulative ideal, by all disputants, in the nature of the situation. Internal disagreement leads to a respect for the truth and if we include as many people as possible inside our reference group, then toleration is possible on a wide scale. In other words, whenever disagreement is internalized it becomes tolerable.

The relativist principle of toleration is therefore a corollary of an unreasonable and antipopulist intolerance. For if I allow every group moral autonomy, it is because I am willing to allow intolerance and cruelty within the group. One warlord will be tolerant of another only so long as the other appears to be strong, and allows the first to do as he wishes within his territory. If I am correct, this model has appeared reasonable to philosophers only because they have neglected study of the importance of dissent (in many different forms) in the generation of the significance of facts, in scientific traditions, and in intellectual traditions generally.

Every society, community, or even institution has its own customs and ways of doing things. This in itself is neither just nor unjust, though it will be a matter of justice when some of these customs will be enforced within the relevant society or community. The question of the justice of the system as a whole must be faced in the intellectual traditions, in which a much more open kind of tolerance is possible than in society. There are, in the intellectual traditions that are available to any community, many viable alternative ethical points of view, some of which will undoubtedly be critical of its customs. In such a situation, justice demands that customs be modified to accede to the criticisms of any viable system of justice that is found in the intellectual traditions. For this reason, no society is ever closed, for it can turn its back on its pluralist intellectual tradition only by embracing some immorality or other as this immorality has been established by standards that are available to it in its own internal intellectual tradition (even if it disingenuously tries to avoid acknowledging them).

In short, my claim is this: Locke contrasted faith with knowledge as the basis of a liberal distinction between what we can and cannot tolerate in society. We can instead contrast faith and the intellectual tradition, which, even if pluralist and sometimes equivocal, is institutionally skeptical, and so acts clearly as a check on what counts as a reasonable or a just act in any community, whatever be the founding myths of the community. The intellectual tradition itself is not above or beyond or greater than our myths, but is itself the social and historical product over many generations of the clash between different myths we have invented: a social product, which, however, is not a social construction, if one understands the word *construction* in its active form. Perhaps we may call it a social by-product, or even a social self-construct, if we think in a general way of this tradition, like most traditions, as forming spontaneously around us as we engage in accordance with it.

To complete this argument, let me go back to Locke's concern in his *Essay Concerning Human Understanding*, where he sets about showing how the skeptical doubt can be avoided. Avoiding it is important for the defence of liberal democracy. Skepticism can be avoided if we pay attention to the distinction between actual and possible controversies in the intellectual tradition. There is no way to confute the Pyrrhonist who argues that given any theory a countertheory could be invented that is just as good. There is no way of confuting the Academic skeptic, who maintains that apparent knowledge obtained by the senses can be reinterpreted in such a way as to make it doubtful as knowledge. A Cartesian skeptic, (i.e., a skeptic who tries to find a foundation for all knowledge, but who fails to find it) may invent a hypothesis of a powerful demon who can make anything appear to us to be true even when is not, and we have no knowledge established thereafter.

Given an actual controversy, I may come to doubt some of the results of even my senses on some point. But to doubt all my senses in every instance I would need to have more than such a controversy; I would need to look at possible controversies, as skeptics do. But then we see where the right answer to the skeptic is to be found: controversies make us doubt controverted statements, and invoke a regulative ideal of truth when we participate in friendly discussion. Participating in a controversy we avow for the sake of the discussion that we do not know the truth, but that we may be mistaken. But this is only for the sake of the discussion, and because the situation demands it. It does not follow that we have any antecedent doubts about it when we are not involved in the discussion (of course, we may have doubts anyway, and of course the controversy may make us doubtful, but not necessarily). We have all kinds of criteria for knowledge with which we are quite satisfied when not challenged, and we do reconsider what falls under them, for the sake of the discussion at least, when challenged. But to consider all possible disagreements at once will of course lead us to say that nothing is truly known, which is the skeptical stance. But we do not ever have all possible controversies before us. We have only those particular controversies that we do have.

The skeptic, moreover, does not raise all possible controversies at once, but raises the mere possibility of all of them. The proper response to skepticism is that we should take controversies one at a time, as they occur, and not, as we may have been tempted to do, to deny all knowledge.[41]

Sufficient unto the day the controversy thereof.

Notes

1. The effectiveness of the critiques, and their effect on scientists, can best be gauged by reading the voluminous debate on the so-called "Sokal Affair" on the Internet, by searching for *The Sokal Affair* using any search engine. This is perhaps the first full-blown intellectual controversy which is felt mostly through Internet exchanges, after the first few exchanges were exclusively in the form of more traditional publications.

2. See Paul Feyerabend, *Science in a Free Society* (London: New Left Books, 1978), and Thomas Kuhn, *The Structure of Scientific Revolutions*, 2nd ed. (Chicago: University of Chicago Press, 1970).

3. Michael Foucault, *The Order of Things: An Archeology of the Human Sciences* (New York: Random House, 1970).

4. See Denis Diderot, *Encyclopedia, or Reasoned Dictionary of the Arts and Sciences* (New York: Harvey Adams, 1978).

5. Jean Le Rond D'Alembert, *Preliminary Discourse to the Encyclopedia* (Indianapolis: Bobbs-Merrill, 1976). The great commotion caused by these books can perhaps be better understood as a reaction to the principle of social contract that pervades them, which claims that all human institutions are humanly made, and thereby denies, by implication wherever it is not explicit, the divinely conferred feudal rights of royalty, nobility, and clergy.

6. We may wish to note that the point of view that the difficulty with the latter part of the revolution was to be found in the character of Napoleon could be endorsed by both the Whigs and the Tories in England, even if they disagreed whether Napoleon's failure was to be traced to his character or to the hopelessness of the quest to liberate Europe.

7. Heine, *Religion and Philosophy in Germany*, trans. Dennis J. Schmidt (Boston: Hugh Mifflin and Co., 1882; Albany, N.Y.: State University of New York Press, 1986), p. 103.

8. Johann Gottfried von Herder, *The Origin of Language*, reprinted in *Herder on Social and Political Culture*, F. M. Barnard, ed. and trans. (Cambridge: Cambridge University Press, 1969), p. 164.

9. Please note that I do not mean by nationalism the rise of nation states in Europe, which precedes the events that we are studying. By "nationalism" I mean the doctrine concerning the moral right of self-determination of every nation.

10. See Elie Kedourie, *Nationalism*, 4th ed. (Oxford: Blackwell, 1993).

11. Herder, op. cit., from Barnard, p. 24.

12. Hegel's theory of classes, clearly the forerunner of those later to become the cornerstone of Marx's analyses, is to be found in his *Philosophy of Right*, ed. and trans. T. M. Knox (Oxford: Oxford University Press, 1942), p. 135. For Hegel's "right-wing" communalism, a better source is his philosophy of history, for instance in the posthumously published (1837) lectures, *Reason in History*, trans. Robert S. Hartman (London: MacMillan, 1953).

13. The effort to rescue politics from the imperialism of reason, as its critics have interpreted this, has led to the worship of art and aesthetics as the true companion of political judgment, and whatever may be its other virtues, it has had the unfortunate effect of supporting unreasonable policy and at the same time mystifying both art and aesthetics.

14. Jagdish Hattiangadi, 'The First World War and 1991', in I. C. Jarvie and N. Laor (eds.), *Critical Rationalism, the Social Sciences and the Humanities* (Dordrecht: Kluwer, 1994).

15. Anil Mashruwala, assistant editor to the *Indian Economics Journal* for many years, was my stern critic on these matters, with whom I had a constant disagreement over thirty-eight years on this very issue. Though not himself a Marxist, but a disappointed Gandhian in the last half of his life, he thought that the theoretical difficulties of Marxism could perhaps be resolved through the right organizational actions, which has seemed to me to be quite impossible in principle, and these doubts have been confirmed in recent history.

16. Heine foresees the great turmoil that was to engulf us in his century and ours, op. cit. pp. 156–161, though he does not foresee that the great communal liberation would be taken up, not just in the lands conquered and then lost by Napoleon, but in the rest of the world as well.

17. An early version of this essay was given as a talk on the occasion of the bicentennial of the French Revolution, and I am grateful to Leslie Green and Peter Danielson, whose incisive comments on that version I have perhaps inadequately taken into account. I wish to thank John Yolton for his careful reading and comment, and Andrew Hill for his detailed textual comments and assistance with this, the final version, of this essay.

18. John Locke, *A Letter Concerning Toleration*, in J. Yolton (ed.), *The Locke Reader: Selections from the Works of John Locke*, (New York: Cambridge University Press, 1977), p. 247.

19. Ibid., p. 270.

20. John Locke, *An Essay Concerning Human Understanding* (New York: New American Library, 1974).

21. Locke's principle of tolerance did not include Roman Catholics probably because the Pope still had nominal claims to all the lands and wealth of the nobility, which had been seized and redistributed by Henry VIII. This makes Locke's work a situated work. Les Green has raised this important point: Locke's writings can be read consistently in this more political and situated manner; the various *Epistles on Tolerance* are epistles by a Protestant Christian to other Protestant Christians. But while all this may be true, there were many Protestant Christians who did not take rationality or modern science as seriously as Locke did. Locke's political philosophy paves the way to an eventually more secular

toleration, which he did not himself endorse, but which is, though situated, a call for reason, rather than merely custom, as the basis of civil law.

22. 'The First World War and 1991', op. cit. See, for a recent interpretation of the terrible events in Kosovo, in the light of history, Noel Malcolm, *Kosovo, A Short History,* which shows with a wealth of detail how the Balkans came to be balkanized with the advent of the "modern" notions of sovereignty, power, and the principle of national self-determination amongst any people.

23. This does not stop a relativist from maintaining that he is a liberal democrat, merely because it robs his claim of all value for political purposes. It is an interesting fact that Jiang Zemin responded exactly in that vein to Clinton in the historic debate concerning the democratic rights of the Chinese people in their press conference on June 27, 1998, before all of China. Indeed, if one does not have standards which are not communally relative, the case can also be made that the United States of America is a socialist country, though in its own way (in fact one can describe any community in any way that seems appropriate).

24. Of course, even if epistemological relativism is false, we could find for moral relativism, and hence against liberal democracy. But I shall postpone for another occasion the study of moral relativism, and address the issue: Is epistemological relativism true? See my 'The First World War and 1991', op. cit.

25. I follow Popper in using the word thus, in contrast to his use of the word *historicism* to denote that other claim of Hegel's, that there are laws of history which determine events.

26. These ideas are borrowed from a short essay read to the World Congress in Logic, Methodology, and Philosophy of Science, in London, Ontario, 1975, as 'After Verisimilitude' cf. the *Proceedings.* That essay was a summary of another entitled 'The Structure of Problems', eventually published in the *Philosophy of the Social Sciences,* 1978–79 in two parts.

27. See *The Open Society and Its Enemies,* and *The Poverty of Historicism* (London: Routledge and Kegan Paul, 1957).

28. *The Poverty of Historicism,* p. 131.

29. Ibid., p. 134.

30. Ibid., p. 88.

31. Those who wish to pursue this question in the context of Popper's theory of maximizing criticism should read Bartley's *The Retreat to Commitment,* where he attempts to rescue Popper from his reliance on faith, in my opinion unsuccessfully.

32. W. V. O. Quine must be given credit for the holistic and naturalistic approach to science, which is also evident in the writings of Feyerabend, and of Kuhn, perhaps as early as Quine's 'Truth By Convention', in P. A. Schilpp (ed.), *The Philosophy of Alfred North Whitehead* (Chicago, Ill.: Open Court, 1937).

33. Feyerabend, *Science in a Free Society,* p. 39.

34. Ibid., p. 39.

35. Ibid., pp. 39–40.

36. Ibid., p. 79.

37. Ibid., pp. 81–82.

38. Ibid., p. 106.

39. Feyerabend's relativism is indeed a fine attack on theories of reason, but as a positive theory it has the defect that he speaks as if he alone falls outside of traditions. This is the incoherent Romantic view that is so easy to fall back upon—that there is a Genius,

Messiah or Leader (or Nietzsche's "Overman") who points the way beyond all traditions to the future.

40. Similarly, though I will not belabor this point, the very idea of postmodernism is incoherent if there are many readings of history—for it presupposes a unique reading of history to find itself defining an era.

41. My thanks are due to Miriam McCormick, whose challenging ideas on skepticism have led me to adopt this position, with which she, as a skeptic, will no doubt not agree.

11

Paul M. Churchland

To Transform the Phenomena

Feyerabend, Proliferation, and Recurrent Neural Networks

Some years ago, I picked out five salient epistemological theses in the writings of Paul Feyerabend, and I attempted to show that those same five theses flowed naturally, and independently, from the emerging discipline of computational neuroscience (Churchland 1996). In short, the claim was that a "connectionist" model of cognitive activity successfully reduces a Feyerabendian philosophy of science. Specifically, the inevitable theory-ladenness of perception, the displaceability of folk psychology, the incommensurability of many theoretical alternatives, the permanent need for proliferating theoretical alternatives, and the need even for proliferating methodologies—all of these modestly radical theses emerge as presumptive lessons of cognitive neurobiology, or so I then claimed. This convergence of principle was arresting to me for all of the obvious reasons, but also because the five theses at issue no longer seemed radical at all, once they were assembled in neurocomputational dress. They seemed harmless and benign, almost reassuring.

I take up the topic again partly because the case now seems to me importantly stronger than it did then. In particular, an additional argument has emerged for the fourth of the five Feyerabendian theses, namely, the methodological virtue of proliferating large-scale theoretical or conceptual *alternatives* to currently dominant theories, even when (especially when) those dominant theories are "empirically adequate". The new argument derives from our increased understanding of the computational capacities and the cognitive profile of what are called *recurrent neural networks*. To outline that argument and explore that profile is the main task of this brief essay.

1. Proliferation: The Original Argument

Feyerabend's argument for the desirability of proliferating competing theories went substantially beyond the widely acknowledged fact that data always underdetermines theory. Given that there is always an infinity of possible explanations for any finite set of data, it remains a permanent possibility that one's current explanatory hypothesis, no matter how successful, is inferior to some alternative hypothesis yet to be articulated. To this much, all may agree. And all may agree that this provides at least some rationale for the proliferation of theories.

But Feyerabend made a further claim. Alternative theories, he argued, have the capacity not just to offer competing and potentially superior explanations of one and the same set of data; they have the unique potential for *transforming* the available data itself, in welcome and revealing ways, and for finding *additional* data where before there seemed nothing but intractable chaos or meaningless noise.

This position is already voiced in Feyerabend's earliest essays. It is a consequence of the fully *theoretical* character of all so-called "observation reports": different theories brought to the very same experimental situation will fund interestingly different families of observation reports (Feyerabend 1958). It is also a consequence of the fact that one's own sense organs, no less than any other physical instruments of measurement or detection, always require some conceptual *interpretation* of the significance of their inevitably complex behavior: distinct theories will not only fund distinct interpretations—they will also select out, as informationally significant in the first place, different elements of that complex behavioral profile (Feyerabend 1962).

Both of these arguments, however, were a fairly hard sell at the time. The first required dumping an entire epistemological theory: classical foundational empiricism. And the second required the adoption of a radically naturalist interpretation of human cognition. Few were moved.

Fortunately, Feyerabend found an arresting way to make the desired point without requiring from his readers a prior commitment to these radical premises. In 'How to Be a Good Empiricist—A Plea for Tolerance in Matters Epistemological' (1963), he cited the case of Brownian motion as a telling illustration of the transformation of intractable chaos and meaningless noise into systematic and revealing empirical data. Brownian motion, you will recall, is the ceaseless zigzag jittering motion of, for example, microscopic pollen grains suspended in water, or smoke particles suspended in air. Its initial discovery in 1827 had no impact on anything, but there was a small puzzle concerning the cause of that ceaseless motion. Robert Brown himself, a gentleman naturalist skilled with the microscope, guessed first that he was seeing the locomotor behavior of tiny animals. But this could not account for the motion of the (clearly nonbiological) smoke particles. Something else was going on, but it was unclear what it might be. The motion of a Brownian particle was an unpredictable rotational and translational meander in three-dimensional space. And it was far from clear that its evidently random dance had any significance at all, even for biology, let alone for physics. It sat more-or-less quietly, without an agreed-upon explanation, for the better part of a century.

As we now understand it, however, the motion of a Brownian particle is one of the most direct experimental manifestations available of both the molecular constitution of matter and the kinetic character of heat. The smoke particles' motion is caused by the individual impacts of the millions of fast-moving air molecules that surround it. The smoke particle is just large enough to be seen in the microscope, yet still small enough to show the effects of the erratic hail of submicroscopic ballistic particles rattling it from every side. Individually, those molecules have motion in that, collectively, they have a temperature above absolute zero. And they continue moving, *ceaselessly*, because they interact with one another in a series of perfectly elastic collisions.

Such a system, remarks Feyerabend, constitutes a perpetual motion machine of the second kind—a system of mechanical interactions in which no kinetic energy is dissipated as heat. The phenomenon of Brownian motion is thus a standing refutation of the classical Second Law of thermodynamics, then one of the best-confirmed "empirical laws" of classical physics. And yet it is far from clear, just from looking at it, that Brownian motion has this giant-killing character. It does not look like much of anything but empty noise. So long as one approaches a Brownian particle with classical thermodynamics in hand, its motion remains an intractable chaos with no clear relevance to anything.

If, however, one approaches the Brownian smoke particle with the quite different assumptions of the kinetic/corpuscular theory of heat, its behavioral profile suddenly teems with potential significance. The permanence of that puzzling motion is only the first of many giveaways. For example, if heat truly is nothing more than molecular motion, then increasing the temperature of the gas should increase that motion, and thus should increase the violence of the smoke particles' jittering dance. Correlatively, lowering the gas's temperature should reduce the violence of that dance. And so it does.

Further, raising (or lowering) the pressure of the gas will raise (or lower) the number of molecules colliding with the smoke particle each second, thus smoothing (or aggravating) the observable character of its motion. And if we hold both temperature and pressure constant, but vary the elemental gas in which the particle is suspended, we should get slightly different Brownian motions for each gas: in hydrogen at S.T.P., a Brownian particle will receive many more (but much gentler) impacts-per-second than it will receive in oxygen at S.T.P., because a hydrogen molecule is only one-sixteenth as massive as an oxygen molecule. Brownian motion should thus be more jagged in oxygen than in hydrogen, and progressively more jagged still in carbon dioxide, krypton, and xenon.

Finally—and most important, historically—if the volume of gas is placed in a gravitational field, then the (much more massive) smoke particles suspended within it will collectively behave like a miniature "atmosphere". On average, each particle will have a translational energy of $3kT/2$, but in fact there will be a statistical distribution around that mean value. This means that some of the smoke particles will have more energy than others, and will thus be found at higher altitudes than others, although the density of such high-flyers will fall off exponentially with increasing altitude, just as with the Earth's atmosphere itself. The experiments of Jean Perrin in 1908, with gum resin particles suspended in a liquid, revealed just such an exponentially fading vertical distribution of those Brownian particles. And from the details of that distribution, he was even able to deduce a credible value for Avogadro's number, a value that cohered well with the values reached by wholly unrelated methods.

Thus the early positive case for the corpuscular/kinetic theory. But let us not forget that all of this novel or reinterpreted data was at the same time a devastating critique of the classical theories of heat, which could not begin to account for any of it; and of the classical Second Law, which asserted the impossibility of the perfectly elastic mechanical interactions that the new theory required, and that the ceaseless jittering of the Brownian particles seemed to reflect.

The point Feyerabend wished to make in recalling all of this was that, without the organizing perspective of the kinetic/corpuscular theory of heat, the subtle information implicit in the broad range of Brownian behaviors would never have emerged from the shadows in which it was buried. Brownian motion, of course, *is* noise. But it is noise with a well-behaved and interestingly variable profile. Brownian motion is a clear historical case where our appreciation of the significance, and even the existence, of robustly repeatable experimental data depended decisively on our approaching the empirical world with a specific theory in mind.

Thus the virtue of proliferating candidate theories in any domain: it occasionally enables us to find a new significance in old data, and to uncover and appreciate new empirical data, data which can then serve to constrain our theoretical beliefs more tightly and more responsibly than would otherwise have been possible. In the end, then, it is not the counsel of a romantic and uncritical pluralism. It is just the reverse: it is a reasoned strategy for enhancing the range of *empirical* criticisms that any theory must face. (Sections 5 and 9 of Feyerabend's 1963 essay make this interpretation perfectly clear).

2. Proliferation: An Old Neurocomputational Argument

My 1996 essay attempted to construct a neurocomputational rationale for the virtue of the Feyerabendian policy of proliferation. If we assume that humans are multilayered neural networks that learn, under the continuing pressure of experience, by the gradual modification of the strengths or "weights" of their myriad synaptic connections, then we should be aware of a common wrinkle in their learning behavior that is evident in even the simplest of artificial models. Specifically, how well a given network manages to learn a certain task (a complex discrimination task, for example) is quite sensitive to the initial configuration of the network's synaptic weights. The same network, subjected to the same training examples, and guided by the same learning algorithm, will often "bottom out", at very different levels of performance, depending on the particular structural/cognitive state it happened to occupy when the training began. As illustrated in Figure 1, the amount of learning (=error reduction) that takes place is *path sensitive*: from some regions in synaptic weight space, the network is doomed to slide into a merely local minimum in the overall error landscape, a minimum from which no series of training examples will ever dislodge it. That is to say, a network can quite properly learn its way into a cognitive configuration that yields nontrivial performance, but then be utterly unable to profit further from continued exposure to the data. The network gets "stuck" in a merely local error minimum. It gets captured, as it were, by a "false paradigm".

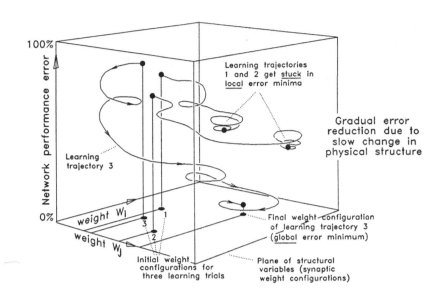

FIGURE 1.

    To maximize the likelihood that we find the weight configuration that yields
the *global* error minimum (that is, *maximal* performance), it is therefore wise to
submit our network to the same training regime several times, starting with a
different *initial* configuration each time. In this way we can escape the blinkers
inevitably imposed by each of a large range of (unfortunate) initial cognitive states.
The lesson for humans, I then suggested, was that we should encourage the proliferation of different intellectual schools, theoretical perspectives, or research programs, in order to achieve the same goal in ourselves.
    The point is legitimate, but I now think the analogy is strained. For one thing,
a single human individual cannot hope to modify his synaptic weight configuration with the ease with which a computer modeler can manipulate an artificial
neural network. An adult human's synaptic weights are not open to casual change
of the kind required. In adults, they mostly do not change at all. Hopping briskly
around in weight space, thence to explore different learning trajectories, is impossible advice to an individual human. My original advice, therefore, can be realistic
only for *groups* of humans, where distinct persons pursue distinct paths.
    Second, I doubt it is true, as the analogy implies, that theoretical progress
consists in the gradual modification of the synaptic weight configurations of the
world's scientists, either individually or collectively. The initial *training* of those
scientists presumably does consist in such a drawn-out process. But the occasional
and dramatic post-training cognitive achievements that we prize seem to involve
a different kinematics, and a different dynamics. Most obviously, the time scales

are wrong. Major scientific discoveries, from Archimedes' "eureka!" through New-ton's falling apple/moon at Woolsthorp, to Roentgen's identification of X-rays, reg-ularly take place on a time scale ranging from a critical few seconds to a critical several days, whereas major synaptic reorganization in the brain takes place on a time scale of months to years. Evidently, humans command a mechanism for conceptual change that is much swifter than the structural reconfigurations that synaptic change involves.

Third, the process of synaptic reconfiguration—to achieve new modes of data processing—entails the simultaneous destruction and abandonment of *old* modes of data processing, the ones embodied in the old synaptic configuration. But we humans are quite able to reconceive data without thereby losing competence in the application of old conceptual frameworks.

These difficulties, and others, would seem to demand a quite different ac-count of the sort of interpretational plasticity displayed in Feyerabend's Brownian motion example, and in other historical examples of the reconceptualization of experimental data. Fortunately, multilayer neural networks with "descending" or "recurrent" axonal pathways provide the elements of such an account. This possi-bility was scouted very briefly (and rather timorously) in one of my essays (1989a). Let me here try to do better.

3. Proliferation: A New Neurocomputational Argument

The contrast between purely feedforward networks and recurrent networks is sche-matically displayed in Figure 2. For our purposes, perhaps the most important difference is as follows. In the purely feedforward case in (a), the network's activa-

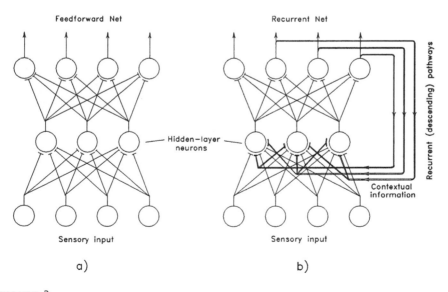

FIGURE 2.

tion-pattern response, across the "hidden" layer neurons, to any pattern of activation levels across the input layer, is dictated entirely by the configuration of synaptic weights that connect those two layers. So long as those weights remain fixed, the same input will always produce the same response (=pattern of activations) at the hidden layer.

In the recurrent case in (b), however, the hidden layer neurons receive axonal input from *two* sources rather than just one. As before, the initial "sensory" layer is one of those sources. But there is also a secondary source of information, a source located higher up in the network's data-processing hierarchy. The cartoon sketch in (b) portrays a recurrent loop that reaches back a single layer, but in multilayer networks such as the human brain, such descending pathways often reach back two, three, or even many layers. (At this point, take a quick look at the puzzling and degraded graphical items shown in Figure 3, and then return to the next paragraph).

Such recurrent information serves to *modulate* the activational behavior of the neurons at the first hidden layer. It can provide a temporary perceptual or interpretational *bias* above and beyond the discriminational biases permanently embodied in the network as a result of its prior training and its relatively fixed configuration of synaptic weights. The recurrent information can slightly "preactivate" the hidden-layer neurons in the direction of a specific, already learned, discriminational category. The result can be that a severely degraded or partial input at the sensory layer will, thanks to the specific character of that recurrent priming, successfully produce the usual or canonical activation pattern (across the hidden-layer neurons) for that learned category. A discrimination that might have been difficult or impossible, in the absence of the recurrent modulation, can be a quick and easy discrimination in its presence. Trained networks tend presumptively to "fill in" missing information in any case—the phenomenon is called *vector com-*

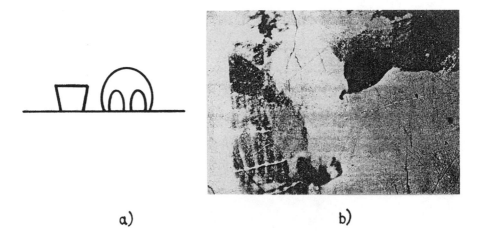

a)   b)

FIGURE 3.

pletion—but recurrent modulation can make a decisive difference between success and failure in any given case.

The graphical items you observed in Figure 3 will serve to illustrate the principle. Quite likely, neither one of them made much sense to you. But observe them again with the following information in mind. (3a) is a portly floor-washer with bucket, seen from behind. (3b) is a close-up of a cow, with its face taking up most of the left half of the picture. A rectangular-grid wire fence can be seen at the lower left, behind the cow, whose shoulder and back lie at the upper-middle right. If you did not see these figures before, you will almost certainly see them now. And seeing them puts you in a position to anticipate the specific ways in which they are likely to behave over time, or in response to changing circumstances. Such behavior, therefore, will display a significance to the prepared observer that would otherwise be absent.

The cognitive activity of a recurrent network, plainly, is a function not just of its physical structure and its current sensory input. It is also a function of the prior *dynamical* state of the entire system. It is a function of the network's prior frame of mind, of the "concerns" it was then addressing, of the "take" it already had on its sensory environment, of the direction of the "attention" it happened then to be paying. We have here a *dynamical* mechanism whereby a network can display, on a very short time scale, quite different cognitive responses to one and the same sensory input. Depending on the contextual information possessed by the network, what gets categorized as chaos or noise in one condition gets recognized as something familiar in another.

A third example will bring out a related point. Observe the scarf-shrouded old woman in Figure 4, with her large bony nose, her chin buried in her woolly shawl, her vacant gaze, and toothless mouth. This is an easy figure and probably you would have come to the same recognition without any help from me. But

FIGURE 4.

now that your visual system is firmly captured by one dynamical attractor or proto-typical category, consider the possibility that you have completely misinterpreted the figure before you, that there is an alternate and much more realistic take on it. Specifically, it is a *young* woman that you are looking at. She is facing away from you, the tip of her nose barely visible beyond her left cheek. What you took to be the left nostril of a large nose is in fact the left jawbone of an entire face, and what you took to be a glazed eye is in fact a perfectly normal ear. Very likely you now see a different organization within Figure 4, and you can probably say what the horizontal slash of the old woman's "mouth" really is.[1]

Once again, recurrent modulation of the activity of your visual system results in a transformation of the scene before you. And in this third case, it represents an escape not just from chaos and confusion, but from a highly compelling alter-native conceptualization, one that might have held you captive, indefinitely, had you been denied the cognitive assistance of an explicit alternative account.

Elsewhere (1989b) I have argued that perceptual recognition is just one spe-cies of explanatory understanding, and that explanatory understanding consists in the activation of one of one's antecedently learned prototype vectors. I call this the *prototype-activation* or P-A model of explanatory understanding, and I com-mend its sundry virtues for your critical evaluation. On this occasion, I wish only to repeat the following claim. Large-scale theoretical or explanatory advances in our scientific understanding typically consist in the fairly swift redeployment and subsequent exploitation, in domains both familiar and novel, of prototypes already learned from prior experience. To cite some familiar examples: planetary motion turns out to be an instance of *projectile motion*, though on an unexpectedly large scale; light turns out to be *transverse waves* of some sort, though of an unprecedent-edly high frequency; and heat turns out to be just the *energy of motion*, although at an unexpectedly small scale.

With recurrent-network models of cognition, we can account for both the initial acquisition of such prototypical categories (by slow learning at the hands of repeated experience and gradual synaptic change), and their occasional and highly profitable redeployment (at the hands of recurrent modulation of our perceptual activity) in domains *outside* the domain of their original acquisition. This two-process model of cognitive activity is represented in Figure 5. In contrast to Figure 1, the learning trajectories here display an occasional discontinuous drop in the error dimension, a drop in performance error that is not owed to any structural (synaptic) changes whatever. The sudden achievement of new levels of perfor-mance is owed entirely to *dynamical* factors that go unrepresented in a typical weight-error diagram. Those dynamical factors consist of the ongoing activational states of the entire network. What is evident from the diagrams, and from the preceding discussion, is that the dynamical system of the human brain can occa-sionally wander into a new region of its vast activational space, a region in which its existing cognitive capacities are put to entirely novel uses, a region in which at least some of its perceptual activities are systematically transformed, thanks to a new regime of recurrent modulation of those activities. Once that regime is in place, one's perceptual exploration of the world can find structure and organiza-tion that would otherwise have remained invisible.

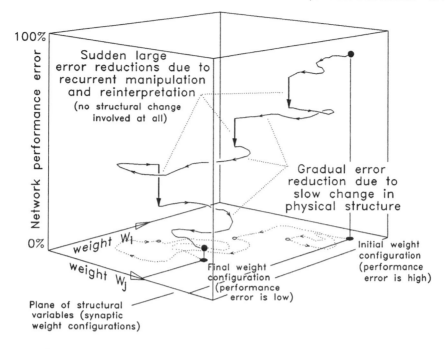

100% — Network performance error

Sudden large error reductions due to recurrent manipulation and reinterpretation (no structural change involved at all)

Gradual error reduction due to slow change in physical structure

0%

weight W_i

Weight W_j

Initial weight configuration (performance error is high)

Final weight configuration (performance error is low)

Plane of structural variables (synaptic weight configurations)

FIGURE 5.

   If this is even roughly how human cognition works, then there is a clear and powerful argument for a policy of trying to bring new and unorthodox theories to bear on any domain where we prize increased understanding: we will not gain the occasional benefits of a systematically transformed and potentially augmented experience if we do not explore the vehicles most likely to provide it, namely alternative theories.

5. Concluding Remarks

Feyerabend, I think, was clearly right in recommending proliferation as a virtuous policy. I hope I have (finally) provided an adequate account, in neurocomputational terms, of why he was right to do so. The politics of administering such a policy, to be sure, remain to be discussed. Thomas Kuhn *shared* Paul Feyerabend's "radical" views on the plasticity and theory-laden character of our perceptual experience, and yet Kuhn recommended a very conservative policy that focuses most of our research resources on the currently dominant paradigm. Kuhn has arguments of his own. For example, a "transformed experience" on the part of a few scattered individuals will not serve to move our collective understanding forward unless there exists a widely shared framework of understanding and evaluation to which such novel "results" can be brought for responsible examination and subsequent dissemination. A shared framework of this kind is of such immense value to the learning process that it must be positively protected from the casual predations of unorthodox enthusiasms, which will always be with us.

I am unable to arbitrate these competing tensions here, and will not try to do so. I confess that my political impulse inclines to Kuhn: it is a chaotic world out there and we must prize the monumental achievements of our institutions of scientific research and education. But my epistemological impulses, and my heart, incline to Feyerabend. Modern science is easily tough enough to tolerate, and even to encourage, the permanent proliferation of competing theories. We need only keep our standards high for evaluating their success. Maximizing the severity of those standards, it is arguable, was Feyerabend's ultimate aim.

Note

1. It is a black velvet "choker" necklace around a slender neck.

References

P. M. Churchland (1989a), 'Learning and Conceptual Change', in A *Neurocomputational Perspective: The Nature of Mind and the Structure of Science* (Cambridge, Mass.: MIT Press), pp. 231–253.

P. M. Churchland (1989b), 'On the Nature of Explanation: A PDP Approach', in A *Neurocomputational Perspective: The Nature of Mind and the Structure of Science* (Cambridge, Mass.: MIT Press), pp. 197–230.

P. M. Churchland (1996) 'A Deeper Unity: Some Feyerabendian Themes in Neurocomputational Form', in R. N. Giere (ed), *Minnesota Studies in the Philosophy of Science, Vol. 15, Cognitive Models of Science* (Minneapolis: University of Minnesota Press), pp. 341–363.

P. K. Feyerabend, (1958), 'An Attempt at a Realistic Interpretation of Experience', *Proceedings of the Aristotelian Society* vol. 58, pp. 143–170. (Reprinted in P. K. Feyerabend (1981), *Realism, Rationalism, and Scientific Method: Philosophical Papers*, Vol. 1 (Cambridge: Cambridge University Press, 1981)).

P. K. Feyerabend, (1962), 'Explanation, Reduction, and Empiricism', in H. Feigl and G. Maxwell (eds.), *Scientific Explanation, Space and Time*, Vol. 3, *Minnesota Studies in the Philosophy of Science* (Minneapolis: University of Minnesota Press), pp. 28–97. (Reprinted in P. K. Feyerabend (1981), *Realism, Rationalism, and Scientific Method: Philosophical Papers*, Vol. 1 (Cambridge: Cambridge University Press, 1981)).

P. K. Feyerabend, (1963), 'How to be a Good Empiricist—A Plea for Tolerance in Matters Epistemological', in B. Baumrin (ed.), *Philosophy of Science: The Delaware Seminar*, Vol. 2. (New York: Interscience Publications), pp. 3–19.

12

Joachim Jung

Paul K. Feyerabend

Last Interview

This interview was conducted in Männedorf Hospital (Zurich Kanton), on January 27, 1994, between 12:45 and 3:15 p.m., two weeks before Feyerabend died. The left side of his body was already entirely paralyzed.

F—Paul Karl Feyerabend
B—Grazia Borrini Feyerabend
J—Joachim Jung

J: How can one promote new ideas in philosophy? And what administrative means would you recommend so that innovations can advance more smoothly? For example in 'Consolations for the Specialist' you wrote: "The normal elements . . . may change because . . . some public figure has changed his mind . . . or because a powerful and nonscientific institution pushes thought in a definite direction" (Feyerabend 1970, p. 214). In Germany, philosophers who bring up new ideas hardly have the opportunity to publish their papers in journals.

F: You mean professional journals.

J: Yes, professional journals.

F: That means: when you say something which is new to the journal you get back immediately: "We cannot publish that". Is that what you mean?

J: Yes. Do you see a way of overcoming this? Because this is a big problem which is being discussed in the German media.

F: By whom? By philosophers?

J: Yes, by philosophers who write for newspapers, like me. I already asked you about this issue in my very first letter.

F: I remember.

J: There you wrote that you cannot say anything about this . . .

159

F: Because I don't know the administrative structure of the philosophy publishing business. I would have to know a little more about that. It seems also in the United States that you send something to a journal and it comes back: "We cannot publish that because of our high standards".

J: And this happens in the States, too?

F: Ja, ja. And also in England if you try to send anything you have written to a journal publisher. Send a book to Oxford University Press, send it to Chicago University Press, and you will get similar reactions.

J: Do you not see a difference between the scientific atmosphere in the States and in Central Europe? In Germany and Austria the United States are considered more progressive, as far as freedom of thought is concerned.

F: Of course, anyone can think what he wants. Publication is the problem.

J: You taught in a lot of countries. I would like to ask you: does the American ideal keep its promises?

F: What is the American ideal?

J: The freedom of ideas, scientific liberality.

F: Ideas are free everywhere. Publication is the problem. Well, look, I taught in Berkeley in California, and I could say whatever I wanted. I was not told: you must follow this, you must use this and that textbook. Usually I found an interesting book I wanted to understand. And the best way of understanding a book is giving a lecture about it. I was completely free in that respect. And it is different in different universities. Also, in Switzerland, when I came here to the ETH [Eidgenössische Technische Hochschule], I was just told: this is a professorship for philosophy of science. I could practically do what I wanted to do.

J: So you had only good experience here in the ETH?

F: Ja, ja, only good experience. Except for the department meetings. The ETH is a great school as far as I was concerned. I don't know whether everybody thinks that. You know, to answer these questions, I would have to know a little more about the philosophy publishing business and also about the administrative structure of universities.

J: What induced you to leave your home country and home town and to work in Anglo-Saxon countries? What is your relationship to Vienna?

F: I don't know. I was born there [but] I don't feel at home [there]. You know, there was a time when I went back and forth between Berlin and London, always: every week three days in London, four days in Berlin teaching. And when I came back to London I felt at home although it was not my mother tongue that I was speaking there. So, it took me a long time before I gave up my position in Berkeley. I ran away from there because I was afraid of an earthquake. I was in the middle of an earthquake in '89. And I said: I don't need this.

J: So you never returned to Vienna in order to work there. Did you have bad experience in Vienna?

F: No, no, no bad experience. [I had an] excellent experience as a student, wonderful. You see: we were all physics students who invaded philosophy lectures and got up in the middle of lectures and said "This is all nonsense, what you are talking about". And we were thrown out of the lectures. There were very good people there.

J: You always emphasize the subjective components of science.

F: The what?

J: Subjective components. You always emphasize that science is influenced by emotions, by feelings, by irrationality.

F: Every human activity is influenced by that. Scientists are just people like anybody else.

J: Yes. Please tell me what were your own motives for dealing with science.

F: Interest. I was fascinated by astronomy. I found astronomical research fascinating, much more fascinating than philosophical research, although one could find any subject in philosophy. Any subject is a good starting point for going into what is called 'philosophy'.

J: You always stressed how important personal experience is for philosophy, i.e., that subjectivity is important for . . .

F: I wouldn't say that, because "subjectivity" is already a philosophical expression which assumes a division between something objective and something subjective. I would never assume that, because these things freely interpenetrate.

J: I would like to know: What were your own motives for dealing with philosophy?

F: Interest. Like somebody who starts playing the piano.

J: Your main claim has been for a variety of methods.

F: That is not a claim. That is just a statement of fact. That is how it is. Just look at the history of the sciences. Compare what some physicists have said at one time and at another time, in some [personal] letters. You find all sorts of methods. And this is not a philosophical position. This is just a statement of fact.

J: But you emphasized this statement again and again. Have you had special experience in this regard? For example were you hindered in . . .

F: No, I was never hindered in anything. No, I was never hindered in anything.

J: In 'Consolations for the Specialist' you wrote that "science . . . is not entirely rational and cannot be entirely rational" (Feyerabend 1970, p. 217). What does rationality mean for you?

F: A set of rules which you are supposed to follow and which says: "If so then it will be this and that", for example: "Always avoid contradictions". This is one important part of rationality. But scientists who avoid that at first try to clear up their theories so that they become contradiction-free. It is not good enough to have a perfectly shaped theory. You have to apply it to different contexts. You see, I was once in Hamburg with Carl Friedrich von Weizsäcker, who invited me to give a talk there. At that time I was still a methodology freak. At that time I believed that it made sense to argue for certain procedures in science. And my arguments were excellent. But von Weizsäcker gave a historical account of the rise of quantum theory and this was much richer and much more rewarding and I realized that what I was talking about was just a dream. Just as Ceaucescu wanted to have order in his country, so he tore down the little houses and built up his concrete monsters. When von Weizsäcker started describing the development of quantum theory he was just pointing out the little houses, because there were so many little steps being made. Niels Bohr said: "When you do research you cannot be tied down by any rule, not even the rule of noncontradiction. One must have complete freedom". So, as he explained that to me, I recognized that my arguments were excellent but that excellent arguments don't count when you want to deal with something which is as rich as nature, or other human beings.

J: Yes, now maybe a little complicated idea. If you claim that all science is imbued by subjective . . .

F: I don't claim that.

J. No, it's a fact. You state this.

F: I will not speak of subjectivity because 'subjectivity' is a philosophical term.

J: Emotionality, or whatever . . .

F: Well, all science always gets input from the entire person who does it and not just from the brain.

J: So you stated that emotions and feelings are components of science.

F: Of scientists doing science.

J: But also of science, of theories. They also contain emotionality if the scientists behave emotionally.

F: Ha, Ha, of course theories are frozen emotions as it were, you know. People have a certain feeling for the world. And then they try to nail their feelings down, and out comes a theory.

J: Yes, in this regard you cannot do one thing: you cannot make an exception of your own theory, your own philosophy . . .

F: I don't have a philosophy. I have opinions.

J: Your worldview, what you taught. This is a philosophy.

F: I didn't teach anything. I told stories. I never gave a systematic account of everything. I was giving lectures, I told stories, you know. So I do not have a philosophy, I have lots of opinions. If you asked me why should I accept [a person's] opinions, I should answer them depending on whom I am talking with, depending on their character, if I know them well enough. So a philosophy is a collection of opinions which are tied together by some general stuff. It's much too rigid for my taste.

J: So I can put off this question because you hinted at another problem. In my lectures and papers I have always regarded Kuhn, Feyerabend, and the Constructivists (Maturana, Varela, von Glasersfeld, Watzlawik) as a unity.

F: No, no, you should have read what I wrote against Kuhn.

J: Yes, I know that you do not agree with Kuhn in every point. I know your remarks on Kuhn.

F: Also I learnt a lot from him. He was my colleague in Berkeley. We have become very good friends.

J: You would not accept that Feyerabend, Kuhn, and the Constructivists constitute a philosophical unity?

F: No, no, Varela doesn't please me at all. I had a big debate with him some time ago at the ETH. Because there we had a seminar where we invited all sorts of people to present their stuff.

J: Yes. What do you think about the ideas of Maturana and Varela?

F: It is not clear what they want.

J: They explain subjectivity in a very reasonable manner. They say that everything is a construction of the subject.

F: You are my construction? That means: if I stopped constructing you, you wouldn't be here any more?

J: No, we are constructed in a similar manner, and therefore you recognize me

F: This means: genetics?

J: Yes, genetics. We are genetically constructed in a similar manner. And therefore you recognize me as a human being and others too.

F: But I also recognize a dog as a canine being. And I am not constructed in the same manner as a dog, except that we are mammals, you know. I recognize frogs, I recognize birds.

J: So you totally disagree with them?

F: At some point it ceases to make sense for me. I also think about and try to figure out what fixed stars have in common with me. So one understands lots of things. But you cannot say: the explanation is a similar genetic code. The whole of astronomy is another example which goes beyond similarity.

`. The point is that your own worldview is in the first place very critical and negative. You have not created a positive philosophical system.

F: In the 'thirties the people said: "There are these analytic strains in philosophy. They just criticize. What we want is something positive to hold us up".

And they called Jewish philosophers 'analytic' because it was the case that many analysts were Jewish. So, you see, I am very suspicious if someone wants something positive. Well, they should create it for themselves. If they want something they should sit down and write a letter to themselves: "My dear friend, I heard you want something positive: here it is . . . "

J: I totally agree with you. It's only because people want a fixed point for orientation. This is why Popper has such a success.

F: He is a founder of a church.

J: Yes, you wrote that.

F: You find positive things not in theories but in human relations. If some-body is happily married, has good children, this is a positive thing in life. It can't be put into concepts. Of course, it can be in some way if a poet writes about it. So there may be positive things and people just don't notice it. They may be married to a very good woman. And even forget that they are married, and not be very kind to her. And children: if children have good parents they get—whatever person it is—a certain amount of self-confidence and also a certain amount of concern for others. And [these] must balance each other. This balance is created by a good upbringing, by good parents who live in harmony with each other. So the positive thing is in the people. What they want is a kind of slogan like "critical rationalism", you know.

J: An ideology.

F: . . . which I can write down and which one can quote. And it's only us who know this critical rationalism, you know. John Stuart Mill, for example, in his *On Liberty*. He says himself, he got it practically from Harriet Taylor, the woman he later married. And he had a good relation with her. Sir Karl didn't have such a good relation with his wife. He ordered her around as if she were a maid. And shouted at her if she brought the wrong book. "Here, you bring me so and so, and it is not the right one!". It was very embarrassing to watch that. So: the positive things are in human relations. They cannot be written down except in poetry or in a theatrical piece. People who want some direction look at the wrong things or the wrong objects to get a statement.

J: Yes. I want to return again to this point. If science is characterized by emotions, too . . .

F: By what?

J: By emotions, by feelings.

F: Not science, scientists are guided by emotions in many cases, very often.

J: So you say that scientists are characterized by emotions and science not at all, that science is pure rationality?

F: Science is nothing. Science is just a heap of words and formulae which have to be filled with content. And it can be filled with content in a different way, namely by trying to make them agree with certain mathematical ideolo-gies which are philosophical, like [that of] the Pythagoreans, or [that] of Einstein, you know. It was filled with metaphysical content because for him nature was a huge [block], he followed Spinoza in this respect, and he was in awe of something independent of him. For him, human beings were aliens on a strange planet. And the scientist tries to find out what the shape of this strange planet is. So all sorts of things appear in a scientific theory.

J: Yes, do you think that emotionality is not reflected in science at all?

F: Where would it be? Where is the emotionality of the second law of thermodynamics?

J: I mean the following: philosophy is sometimes considered a science but there are a lot of different worldviews. You taught that in physics it's the same. But in philosophy the differences between the single views are larger than in physics and in the exact sciences.

F: You mean that the difference between Kantians and Heideggerians . . .

J: Yes, for example.

F: . . . is larger than the difference between Einstein and Niels Bohr?

J: I think so . . . You do not like the concept of 'Wissenschaftstheorie' (philosophy of science).

F: I do not like the *business* of 'Wissenschaftstheorie'.

J: In *Against Method* you wrote: "Bastard subjects such as the philosophy of science which have not a single discovery to their credit profit from the boom of the sciences"[1] Also, on other occasions you wrote that you do not like this concept of "Wissenschaftstheorie". What I wonder is: what do you do instead of 'Wissenschaftstheorie'?

F: What do I do?

J: What is your profession?

F: My profession was: I was a professor of philosophy. This means: a civil servant, ein Denkbeamter. That's all I was. What do I do? I write essays which upset people.

J: You plead for publishing and discussing every idea in general but what about the so called 'Auschwitz lie'? What about race theories? Do you think that adherents of totalitarian ideologies have the right to deny historical facts and to exercise hate propaganda?

F: Hate is never a right. Hate is never a right.

J: So you think that these thoughts should be suppressed?

F: Suppressed? One should be very careful because people should know that such thoughts exist. Then they will be prepared to face the consequences. If you suppress them they start festering somewhere and nobody knows about them and suddenly it explodes . . .

J: So you think one should discuss about everything and . . .

F: Well, discussing means clarifying and making clear what the dangers are.

J: Do you think that tricky theories should not be suppressed, that anti-Semitism too should be . . .

F: Anti-Semitism is not a theory, it's an attitude.

J: An attitude, yes, . . . should be discussed also from the point of view of anti-Semites?

F: I have seen some of this stuff here, you know, the skinheads and the German right. Look, I followed this attack on the mayor of Vienna which surely made me angry. But people should know about it. That is the reason why it should be discussed. People should know that in their midst there are a few very weird people. Because, after all, some of them may teach at school.

J: I do not know whether you heard about this problem too, that a few years ago the Wittgenstein Symposium was canceled because Peter Singer had been invited. There were demonstrations, and then they canceled the symposium because some people did not like Singer coming to Austria. Was that a right decision? How do you judge on this?

F: It really depends what the consequences were. For example: Peter Singer came to do a seminar in Zurich. He was thrown to the ground and his glasses were stepped on. It should be considered, you know. It is like abortion, you know. It is a similar situation as in the case of abortion. People should learn about his views. Because he is somebody who makes suggestions which at first look very plausible. For example: people are completely disfigured. He says: "Life would be a burden for them, so maybe they should not even come into the world. So, there should be full abortion rights". This is something you have to think about. Also, if somebody is in hospital and is barely kept alive with the latest technology, this can go on forever. Should not one rather pull [the plug]? That is something you should think about because there is death in families. You see: earlier on it was different. Because earlier on the family always considered disfigurement and death as day-to-day events. And people were forced to think about things like death and disfigurement.

J: In 'Consolations for the Specialist' you write: "It seems to me that the happiness and the full development of an individual human being is now as ever the highest possible value" (Feyerabend 1970, p. 210). If one can generalize . . .

F: You see, I would restrict this. Not only the full development of human beings as 'individuals'. Human beings are gregarious. One should also take into account 'related happiness', happiness [of human beings] in relation to others. You know: this stuff like Singer, it is important that people think about it. Because they will be faced with decisions of this kind throughout their life.

J: If this statement can be generalized, one could say: if happiness and full development is the highest possible value you consider ethics the highest philosophical discipline.

F: What?

J: Ethics. This means that ethics is more valuable than epistemology . . .

F: Much more so. Morals is much more valuable than just abstract knowledge because morals come from a long development process of people adjusting to each other.

B: But you get this: it's morals, not ethics.

F: Ethics is the theory of morals. It is too philosophical for my taste. Morals depends on time, on history. It is more important than abstract knowledge like ethics, stuff like "What is 'the good'?"

J: Why, I wonder, did you never or very seldom discuss questions of ethics, of morals?

F: Because most of the stuff I read bored me to tears . . . But matters of politics and knowledge—which I often talked about—these are moral questions. Whether knowledge gained by a group should be regarded as objective and therefore imposed on everybody. That is a moral question. Also, what should be taught in schools is a moral question. What should be propagated all over the world is a moral question. What should be taught and discussed in schools . . . Should [the ideas of] Peter Singer be discussed in biology or in history [courses]? So that people [become] aware of their dangerous pitfalls.

J: I think that ethics can be determined less rationally than the other disciplines of philosophy. Would you accept the view that ethics contains less rationality than . . .

F: Fortunately! Fortunately. . . . You can acquire a lot of rationality. Just read Kant, *Grundlegung zur Metaphysik der Sitten*; that's really a rational ethics.

J: Do you mean . . .

F: You can twist everything into a rational shape. Even poems can be twisted. Just read the literary critics [on] how to write poems, that is a kind of theoretical shape. Anything can be bent in a direction, so ethics certainly can.

J: So you think in general that one can give reasons for ethical actions?

F: For actions? Certainly.

J: How would you describe the relation between ethics and rationality?

F: Well, rationality is an attitude. Rationality is an attitude, you see. To approach something rationally is to approach it with a certain attitude. The quest for clarity is a certain attitude; [in that case] numbers are better than words. It started with Plato. It is an emotional attitude, the yearning for shining clarity, like a disinfected bathroom. And, of course, scientists are very competitive and very emotional in their own domain: see what goes on just before the Nobel Prizes are distributed, there is a lot of lobbying, they spread bad rumors around, etc.

J: (Question about Viennese dialect)

F: When I was young I disliked it but this has now gone away completely. In Zurich I loved to speak it: the Vienna-deutsch . . . I had all sorts of strange attitudes in my life.

J: (Question about Berkeley as 'intellectual desert')

F: Yes, especially when I was coming from London, which is full of theaters . . . We were saying before that rationality is an attitude. If you read Einstein you find that his objective reality of course does not have in it what one usually calls 'emotions', but the search for an objective reality is driven by strong emotional forces, and that begins with Parmenides, who said the world is one simple block which does not change and has no subdivisions. . . . In the United States there are many people who are much better than their so called superiors. I was very fortunate. Now I could be stuck in Vienna as a salesman in a bookshop (which would not have been so bad: I would have read the books). What does not please me is to see some idiots getting large amounts of money in important positions [while] some smart young people [are] being pushed around, with no jobs, no money, nothing.

Note

1. *Against Method*, (London: New Left Books, 1975), p. 301.

References

P. K. Feyerabend (1970) 'Consolations for the Specialist', in I. Lakatos & A. Musgrave (eds.), *Criticism and the Growth of Knowledge*, (Cambridge: Cambridge University Press), pp. 197–230.

Index